ROUTLEDGE LIBRARY
SOCIAL AND CULTURAL (

Volume 4

THE GEOGRAPHY OF THE
NATIONAL HEALTH

THE GEOGRAPHY OF THE NATIONAL HEALTH
An Essay in Welfare Geography

JOHN EYLES

Routledge
Taylor & Francis Group

LONDON AND NEW YORK

First published in 1987

This edition first published in 2014
by Routledge
2 Park Square, Milton Park, Abingdon, Oxfordshire OX14 4RN

and by Routledge
711 Third Avenue, New York, NY 10017

First issued in paperback 2016

Routledge is an imprint of the Taylor & Francis Group, an informa business

British Library Cataloguing in Publication Data
A catalogue record for this book is available from the British Library

ISBN: 978-0-415-83447-6 (Set)

ISBN 13: 978-1-138-98937-5 (pbk)
ISBN 13: 978-0-415-73156-0 (hbk)

Publisher's Note
The publisher has gone to great lengths to ensure the quality of this reprint but points out that some imperfections in the original copies may be apparent.

Disclaimer
The publisher has made every effort to trace copyright holders and would welcome correspondence from those they have been unable to trace.

THE GEOGRAPHY OF THE NATIONAL HEALTH

AN ESSAY IN WELFARE GEOGRAPHY

JOHN EYLES

CROOM HELM
London • Sydney • Wolfeboro, New Hampshire

Croom Helm Ltd, Provident House, Burrell Row,
Beckenham, Kent, BR3 1AT
Croom Helm Australia, 44-50 Waterloo Road,
North Ryde, 2113, New South Wales

British Library Cataloguing in Publication Data

Eyles, John
    The geography of the national health: an
    essay in welfare geography.
    1. Medical geography — Great Britain
    2. Medical care — Great Britain
    I. Title
    362.1'0941   RA847

    ISBN 0-7099-0928-4

Croom Helm, 27 South Main Street,
Wolfeboro, New Hampshire 03894-2069, USA

Library of Congress Cataloging-in-Publication Data

Eyles, John
    The geography of the national health.

    Bibliography: p.
    Includes index.
    1. Medical care — Great Britain — Utilization.
2. Medically underserved areas — Great Britain.
3. Resource allocation. 4. Public welfare — Great
Britain. I. Title [DNLM: 1. Delivery of Health Care —
Great Britain. 2. Medically Underserved Area — Great
Britain. 3. Public Policy — Great Britain. 4. Social
Welfare — Great Britain. W 84 FA1 E9g]
RA410.9.G7E95   1987   362.1'0425   86-29120
ISBN 0-7099-0928-4

Printed and bound in Great Britain by
Biddles Ltd, Guildford and King's Lynn

CONTENTS

Foreword
Preface

To David
with love

Foreword

This book about the geography of the national
health, as viewed as outcome of resource allocations
themselves based on power relations and conceptions
of need and justice, has been a long time in the
making. It began as a joint project between John
Eyles and kevin Woods and has ended as a solo effort
by the former as the latter left Queen Mary College
to take up an appointment in the NHS. Delays were
caused by trying to retain the book as a joint
venture but because of kevin Wood's new commitments
this became increasingly to mean that it would
perhaps never see the light of day. Generously,
Kevin Woods has allowed his work to be used and
modified. I gratefully acknowledge the use of that
material. The necessary take-over of the volume by
John Eyles has perhaps meant that emphases have
changed. Thinking about and drafting the conceptual
chapters and then trying to relate them to empirical
and policy matters has led to the conclusion that
this book is a start to reconstruct, in a particular
way, the geography of welfare: hence its sub-title.
This may lead health care geographers to view the
volume unfavourably, although in its defence it may
be suggested that health and welfare themes from the
new Poor Law to the 1980s are used to illustrate the
broader themes, which of necessity go beyond 'geography of the national health'. And to be meaningful, a geography of welfare must treat not only the
where question but also the who, the what and crucially the how. It is hoped that these themes may
be further tied together in a third (and final)
volume on comparing health and welfare systems.
    Some acknowledgements are in order: Carol Gray
yet again produced high quality camera-ready copy,
Leslie Milne excellent maps and diagrams and Kevin

Woods, especially early on, shared his insights. Peter Sowden at Croom Helm must be thanked for letting many final deadlines pass. My family put up with yet another absence, this being particularly hard on a three-year old. And it is to him that I dedicate this book.

Preface

The purpose of this book is to describe and analyse the geography of the national health with special reference to Britain. It is in many ways a study in welfare geography as its central themes are covered by the question - who gets what where and how. The various parts of this question structure the book and form the foci of different chapters. The 'what' is taken up in Chapter One as we establish a broad-based view of health, care and resources. Our approach is not only broad-based but also integrated seeing health as well-being and hence related to housing, welfare, employment, environment and so on. This integrated theme is taken up in Chapter Four after the initial addressing of the questions of how and who. The 'how' is seen in terms of the mechanisms that allocate care and resources to different individuals, groups and territories, and is treated in Chapter Two. Chapter Three examines the 'who' with respect to the social and epidemiological bases of health and illness in Britain. The 'where' is a theme running through the opening chapters and it is explicitly addressed as people in space (regions and localities) in Chapter Three and allocation across space (spatial patterning of resource allocations) in Chapter Two. The 'who' dimension is further taken in Chapter Four which explores, as has been stated, the integrated nature of health as well-being. Chapter Five investigates the policy implications of an integrated approach. The chapter is also at pains to point to what policy does not do (but what models imply) as well as what is achieved. The final chapter, six, summarises the main arguments of the book, comments on the nature of welfare geography, assesses the impact of integrated approaches on the policy process and points the way forward to geographies rather than a geography of the national health.

Chapter One

HEALTH, CARE, RESOURCES

Introduction

The purpose of this book is to examine the shape and
texture of the geography of the national health. We
intend, however, to take a specific view of 'geography' and 'national health'. By geography, we mean
not only the patterning of social and epidemiological conditions across territories. We also
regard the spatial basis of policy initiatives to
tackle problems emanating from these conditions as
geography. A geographical perspective is, therefore, a necessary frame of reference for any analysis of health and health care. It is not of course
a perspective sufficient to itself. Any geography
must be part and parcel of a broad-based examination
of health. A fuller discussion of this view may be
found in Eyles and Woods (1983). The term 'national
health' is derived from the British experience and
may be regarded as an implicit reference to the
national health service (NHS). We do not, however,
intend to present a discussion of the changing
nature of the NHS. There are several recent exposi-
tions of such developments (Allsop, 1984; Klein,
1983). Rather, we are concerned with the nature of
'health'. Indeed, this chapter will deal with the
conceptual problems of 'health' and 'illness' - the
what question - and we shall see that the 'national
health' in essence means the national illnesses and
that the NHS would be better called the NIS (the
national illness service) or the NSS (the national
symptom service). As may be surmised 'national' is
simply taken to mean health and health care in one
country, specifically, in our case for the purposes
of this book, Britain. We recognise that illness
and disease can be 'exported' as in the occupational
hazards of branch plants and factories in the Third

1

World and the iatrogenic illnesses associated with
drug exporting policies in the same regions of the
world. We must also note that styles of health care
delivery in one part of the world will influence,
affect and shape those in other areas as has happened with the US private system of provision
influencing British policy and practice and the
Australian system of funding health care being based
on the Canadian. These interrelations are, however,
beyond the scope of this book.

There are other interrelations which are at the
root of this work. In our discussion of the geography of the national health, we are concerned with
the relationships between the geographical differentiation of health states and the geographical basis
of policy, between the nature and significance of
health and other dimensions of people's lives, and
between the shape of health care policy and other
government policies which materially affect the way
people live. We are concerned that, particularly at
the policy level, there is a lack of recognition
that these interrelationships exist or are important. Much of this book will, therefore, be taken up
with documenting and assessing these relationships
and at the failure of policy to address them. In
some respects, it is difficult to assess something
that does not happen. But it is of course just as
important to recognise what policy does not do as to
look at what it does. The policy option of an
integrated health and welfare system is available
but has not been established. Indeed, we think that
such an integrated approach becomes less rather than
more likely. But we also see the need for such an
approach becoming increasingly significant as more
expenditure on 'health care' fails to greatly
improve the nation's quality of health.

The book can be seen as being organised round
these rather nebulous themes, as well as those of
'welfare geography' established in the preface.
Chapter Two explores the allocational bases of
health care - the how question - demonstrating that
society may treat 'need' in particular ways and that
this treatment is predicated on a particular view of
social justice. In turn, this notion of justice
will shape the systems and mechanisms of resource
and care allocations to determine who gets what.
Chapter Three deals explicitly with the who
question. Chapter Four will, however, take this
question and the integrated approach further and
examine explicitly health in relation to housing,
work, environment and so on, while Chapter Five will

look at actual resource allocations between different policy arenas and within the health budget with
respect to public expenditure. It also examines
regional and local allocations in the light of
specific spatially delimited problems. Chapter Six
discusses finally the prospects for welfare geography and the national health. As, however, has
been suggested, all these examinations assume a
particular view of health. Indeed, they also assume
specific treatments of care and resources. Thus in
the remainder of this chapter, we shall discuss
health, care and resources, contrasting our view
with those usually adopted in policy.

Health

While we talk of 'health' and 'health care', it is
perhaps more accurate to use the terms 'lack of
illness' and 'medical care'. It is of course now
well-known that health and health care are today not
only shaped but also dominated by medical definitions and practices. We see health through the
disease-oriented approach of the medical model in
which the treatment of symptoms and curative processes dominate (Eyles and Woods, 1983). Medical
definitions and practices thus dominate what Rein
(1976, 103) has called the 'policy paradigm' of
health care, a policy paradigm being a "curious
mixture of psychological assumptions, scientific
concepts, value commitments, social aspirations,
personal beliefs and administrative constraints."
The rise of scientific medicine and professional and
popular expectations of medicine have resulted in a
treatment- and curative-oriented health care system
within the British NHS.
    Medicine does not only dominate the policy arena
but is said to influence significantly the nature of
social life as well. Zola (1972) writes of the
medicalisation of social life. He sees medicine as
becoming the new repository of truth, the place
where absolute and often final judgements are made
by supposedly morally neutral and objective experts.
More and more of people's lives are opened up to
medicine as doctors feel that they need to know
about the nature of daily living as well as bodily
symptoms. Medicine also retains near monopoly
access to 'taboo' areas, such as the inner workings
of body and mind. This monopoly is such that many,
if not the majority of, lay theories about the way
the body works are at odds with the scientific
theory. For example, in her study of Britons of

3

Afro-Caribbean descent, Donovan (1965) demonstrated
how the importance of expected flow (related, for
example, to the digestive system, menstrual cycle
and blood circulation system) affected perceptions
of health and of treatment. Liquids were seen as
the best form of treatment as they would flow
straight to the source of discomfort. Pills, on the
other hand, took longer as they had to be absorbed.
Further, medicine retains absolute control over
certain technical procedures and at the same time
has expanded its influence to define what is relevant to the good practice of life itself. As
Freidson (1970) notes medicine has first claim to
that which falls under the description of illness
and anything to which it may be attached, irrespective of its capacity to deal with it effectively.
Medicine thus forms what Illich (1977) called a
'radical monopoly', an institution forever extending
its sphere of competence, prestige and domination.
In turn, Illich himself sees medicine as actually
inducing particular disorders and adverse conditions
among the population through various dimensions of
the process of iatrogenesis: "clinical, when pain,
sickness and death result from the provision of
medical care; ... social, when health policies
reinforce an industrial organisation which generates
dependency and ill-health; and ... structural, when
medically-sponsored behaviour and delusions restrict
the vital autonomy of people by undermining their
competence in growing up, caring for each other and
ageing" (Illich, 1975, 165).
    Illich, perhaps rightly, sees greater individual
responsibility for health and health care as a
necessary corrective to iatrogenesis. He does,
however, overstate his case in that medical care is
a necessary and often beneficial dimension of health
care. Much has been done to put the achievements of
medicine into the context of, for example, the
environmental improvements that did much to remove
the insanitary conditions which assisted the spread
of infectious diseases (see Powles, 1973; McKeown
1979). Further, medical procedures are often less
effective than once thought after proper trials.
Cochrane (1972) cites oral therapy in the treatment
of mature diabetes and the surgical treatment of
some forms of cancer as examples, while studies by
Mather et al (1976) and Hill et al (1970) point out
that hospital treatment does not convey any significant benefit over home circumstances and GP care for
cases of uncomplicated myocardial infarction. But,
in other instances medicine and medical care have

literally changed the world as both Dixon (1978) and Dollery (1978) point out with respect to antibacterial chemotherapy. Morris (1980) draws attention to the role of treatment for those chronic conditions in which effective prevention is still not possible. And in a review of fifty years of medical therapy Beeson (1980) concludes: "a patient today is likely to be treated more effectively, to be returned to normal activity more quickly, and to have a better chance of survival than fifty years ago. These advances are independent of such factors as better housing, better nutrition, or health education."

There is, therefore, no denying the significance of medical treatment and intervention in daily lives. But just as its influence is not detrimental, neither is it all beneficial. For our present purposes, its potentially detrimental and limiting effects emanate from its definition of health care and its role in the social order. The domination of medicine by a scientific, mechanistic and symptomatic view of its practice is well-documented (see Dijksterhuis, 1961; Reiser, 1978; Rossdale, 1965; Doyal, 1979). The domination of this view has resulted in the virtual banishment of 'unscientific' methods and treatments and its emergence has been portrayed as a change from bedside to hospital and laboratory medicine (Jewson, 1974; 1976). In the former, the individual was treated as a whole person to be restored to 'health' whereas in the latter disease, symptoms and active medical intervention to effect a cure of the faulty parts of an individual organism are emphasised.

This interventionist view is extremely persuasive because of the power and prestige of medicine in society. Medicine treats ourselves - our bodies and minds - in other words, that which is most important to us. Its importance rests on its centrality to our lives and life itself. Because of this centrality, its definition of health is often unquestioned or taken-for-granted in everyday life. The fact that doctors are part of the social elite is of course important but inconsequential for purposes of this argument. It is the failure to question their definitional monopoly (which emanates in part from their societal position) that leads to a view of health as a residual category, as the way we are and feel after suitable prescribed treatments and not how we could feel. Alford (1975, 195) is, therefore, correct when he states "physicians have extracted an arbitrary set from an array of skills and knowledge relevant to the maintenance of health in a population and have successfully sold these as their property for a price and have managed to create legal mechanisms which enforce that monopoly and the social beliefs which mystify that population about the appropriateness and desirability of that monopoly."

It is, however, 'medicine' rather than 'physicians', the institutional and structural parameters rather than individual practices, that is important. The effects of this definitional monopoly are enormous, and as the improvements in health resulting from better nutrition and living conditions occurred at the same time as the extension of therapies and drugs available to medical practice, these improvements are subsumed under the medical definition. In Taylor's (1979) view, its effect has been the 'patientisation of the population' whereby people rely on medical intervention for all manner of problems and expect medicine to so intervene. In this view, Taylor is much influenced by Thomas (1975) who argues that people are educated to view themselves as fundamentally fragile, always on the verge of mortal disease and perpetually in need of support by health professionals. "The trouble is", he says, "We are being taken in by the propaganda, and it is not only bad for the spirit of society, it will make any health care system, no matter how large and efficient, unworkable." Herein lies the problem of trying to create 'health' by the removal and treatment of illness and disease. As was found immediately after the initiation of the British NHS, demand is insatiable. Fuchs (1974, 4) regards it as hardly news that "we cannot all have everything that we would like to have". And he is right to implicate the parsimony of nature in this regard. But he is wrong in suggesting that this basic human condition cannot be attributed to the 'system'. The parameters may be sent by nature but the allocation of resources is a fundamentally human and political exercise in which definitions of need and justice play a significant role (see Chapter Two). For health care allocations, the definition of health is of course crucial. A residual definition of health is extremely problematic as it is static approach to a dynamic process. It suggests that health is a goal, indeed an achievable goal, rather than something which may be lost or which fluctuates. 'Health' in the medical definition is not only residual but also progressive, assuming that we can, for our age, gender, place of residence, achieve or

6

progress to a state of health. This is, as Dubos
(1959, 221) points out an ideal dream. "Man (sic)
cannot hope to find another paradise on earth
because paradise is a static concept while human
life is a dynamic process". Health must, therefore,
be seen in dynamic interrelation with an individual's life and the circumstances within which that
life is lived.

Medicine thus defines health as the relative
lack of illness and health care as the curative
treatment of ailments afflicting individual organisms. In a way, health care policy could well be
labelled illness cure interventions because of the
power of medical definitions. But how else might
health be treated? It is in fact difficult to find
evidence for a 'non-medical' view of health in the
studies that have examined individual behaviour and
lay concepts of health and illness. Indeed, most
studies of behaviour are of illness behaviour, ie
the processes by which individuals become aware of
health problems and the identification of types of
treatment and referral patterns which such perceptions engender (see, for example, Mechanic, 1978,
Dingwall, 1976). Studies of lay conceptions in
industrial societies have been few (eg Herzlich,
1973; Blaxter and Paterson, 1982; Cornwell, 1984,
Pill and Stott, 1982; Williams, 1983). Their main
conceptions are indeed medical as people share a
common, medicalised culture. Cornwell (1984) notes
that while her respondents had public accounts -
sets of meanings in common social currency which
reproduce and legitimise the assumptions people
take-for-granted about the nature of social reality
- of health, they only had private accounts - those
springing directly from personal experience and from
the thoughts and feelings accompanying it - of
illness. But what this work on lay conceptions
points to is that health must have always been in
relation to other ideas and dimensions of life. It
is seldom viewed by individuals as an isolated
phenomenon. It is possible to regard Parsons'
(1951) much criticised work on the sick role as the
instigator on this social, relational conception.
The characteristics of this role are that the individual is not held responsible for his/her condition
and is exempted from normal responsibilities as long
as s/he desires to get well and seeks competent
help. The critique of Parsons need not delay us
(see, for example, Ehrenreich and Ehrenreich (1978)
who point to the central absence in the notion of
who and/or what determines the social construction

7

of sickness), because by inference we can see health defined in terms of normal social functioning. If a person is able to fulfil her/his social obligations and duties, s/he is healthy. This usually means functioning with a number of what are medically identifiable symptoms, but Parsons is perhaps close to our commonsense notion of health, that suggests a normal state of affairs and a feeling of well-being (see also Dingwall, 1976; Brearley, 1978).

But in many ways, these feelings are unique to the individual. Nader and Maretzki (1973) suggest that health is closely related to the ways in which individuals construct their social worlds. There are, however, commonalities in the 'materials' used in these constructions. Age, gender, class, locality are all more or less shared conditions which are implicated in this sense of well-being. As Cornwell (1984, 145) comments "it is not enough to know that health is interpreted as 'functional ability' or 'capacity to work' ... without knowing something about the nature of the work people do and of their relation to it." It is, however, on another level sufficient to know of the relation between health and work (or the lack of it), because it is only one small step from this relation to say if health is in relation then health care policy must also be implemented in relation to other crucial arenas.

There is then something elemental rather than utopian about the World Health Organisation's (1961) definition of health as "a state of complete, physical, mental and social well-being and not merely the absence of disease or infirmity." It is utopian to suggest that individuals can be so free of disease and infirmity as to enjoy complete well-being. It is elemental in the sense that 'health' has come to mean for most people the ways that they feel about their lives and life in general. This should not be regarded as the medicalisation of all dimensions of life but with regard to medicine its expropriation of 'health' and, as we have seen, its particularistic definition and use. This expropriation has been made comparatively easy by the taken-for-granted nature of health as well-being. Indeed, the salience of health as well-being is demonstrated only by disruptions to its taken-for-grantedness, this lying in language itself. This disruption can be exemplified by some of Garfinkel's experiments with his students.

(S) how are you?
(E) How am I in relation to what? My
    health, my finances, my school-work,
    my peace of mind, my ...?
(S) (Red in the face and suddenly out of
    control) Look! I was just trying to
    be polite. Frankly I don't give a
    damn how you are.
My friend and I were talking about a man
whose overbearing attitude annoyed us. My
friend expressed his feeling.
(S) I'm sick of him.
(E) Would you explain what is wrong with
    you that you are sick?
(S) Are you kidding me? You know what I
    mean.
(E) Please explain your ailment.
(S) (He listened to me with a puzzled
    look) What has come over you? We
    never talk this way, do we?
                           (Garfinkel, 1967, 44).
    Further demonstration of its salience can be
found in more formal surveys than this ethnomethodo-
logical investigation. Subjective social indicators
are devices for measuring an individual's level of
satisfaction-dissatisfaction with particular
characteristics of life (or life-domains). Individuals are usually asked to rank all or the most and
least important of these domains, which include such
dimensions as standard of living, job, family life,
health, marriage, house, friends, spare time, town
and district, law and order and welfare services.
In samples drawn from the nation-wide population of
Britain (Abrams, 1973; Hall, 1976) as well as in
particular localities of varying type, i.e. villages
(Ward, forthcoming), small-towns (Eyles, 1985a) and
cities (Hunt, 1985), health was ranked as the most
important life-domain. The reason for this is that
'health' encapsulates so many different dimensions
of well- and ill-being. It constitutes the ways we
feel not only physically but also psychologically
and emotionally. And it may be argued that 'health'
contains social and economic dimensions as well as
work; standard of life and social life are significant aspects of our quality of life which is often
described as 'health'.
    Such a broad conception of health does not make
it meaningless. Indeed, it is possible to argue
that its narrowing over the past two to three hund-
red years by scientific medicine has resulted in a
mechanistic, curative approach to care. In some

ways, the narrowing was necessary but its consequences have been unfortunate. "If 'sickness' and
'health' were to lay claim to public resources, then
these concepts had to be made operational. Ailments
had to be turned into objective diseases. Species
had to be clinically defined and verified so that
officials could fit them into wards, records, and
budgets and museums. The object of medical treatment as defined by a new, though submerged, political ideology acquired the status of an entity that
existed quite separately from both doctor and
patient" (Illich, 1975, 111). Medicine arrived and
'health' departed.
    Support for broad conception can also be derived
from social theory, particularly the work of Simmel
(1950), who regarded the 'metropolis', or urban
society, as cultivating a particular form of 'mental
life', one which would allow the personality "to
withdraw from work and to become based upon itself"
(Simmel, 1950, 275). Urban life allows the individual to think and worry about whom s/he is -
'health' - but not in conditions of his/her choos-
ing. The possibilities for creating and enhancing a
sense of well-being are present but life is composed
of impersonal and objectified contents so as the
subjective social indicator studies show, we regard
our 'health' as being of paramount importance but
seldom find its our most satisfying life-domain.
"The deepest problems of modern life derive from the
claim of the individual to preserve the autonomy and
individuality of his existence in the face of overwhelming social forces, of historical heritage, of
external culture, and of the technique of life "
(Simmel, 1950, 409). Specialisation and its concomitant social and economic interdependence limit
the potentialities of urban life for the creation of
whole human beings. 'Mental life' is dominated by
impersonality, isolation and alienation. With such
a low level of psychological well-being, a great
strain is placed on physical health and functional
ability and capacity. The forms of health are
conjoined and become problematic in an urban
society. As Simmel (1950, 422) comments
        The individual has become a mere cog in an
        enormous organisation of things and powers
        which tear from his hands all progress,
        spirituality, and value in order to trans-
        form them from their subjective form into
        the form of a purely objective life. It
        needs merely to be pointed out that the
        metropolis is the genuine arena of this

culture which outgrows all personal life.
Here in buildings and educational institutions, in the wonders and comforts of
space-conquering technology, in the forma-
tions of community life, and in the visible
institutions of the state, is offered such
an overwhelming fullness of crystallised
and impersonalised spirit that the personality, so to speak, cannot maintain itself
under its impact.
The context of life appears rich and full of opportunities but these do not appear to be realised in
life itself. The social order provides but, by the
way of its provision, immediately removes the possibilities for personal fulfillment, for a complete
and harmonious mental life. Of course, urban
society is not alienating in its entirety, but
consideration of Simmel allows for a broad conception of health (as well-being) and a location of
that conception in the social order itself. In the
final analysis, therefore, what constitutes health
must be seen in relation to its societal context in
terms of definitions and structures.

A full discussion of this view is beyond the
scope of this book (see Eyles, and Woods, 1983,
Chapter Six) but it will be drawn on when we examine
the context of health care policy in Britain.
Societal context determines which policies are
instigated, supported, rejected or never implemented. Further, this broad view of health has implications for the types of policies that can be legitimately addressed as pertaining to health and
health care. As we shall show, we regard, amongst
others, housing, social services, environmental and
employment policies as relevant. A broad view of
health also demands similarly broad conceptions of
care and resources. If health is seen as the
meaningful summary expression which people use to
talk of their well-being then it seems right to
acknowledge that responsibility for ensuring as high
a level of health as possible is both individual and
collective. We are not suggesting that one form of
responsibility is more important or better than the
other. Much will depend on 'needs' and circumstances as well as the level of 'resources' available to enhance health. As we shall see,
'resources' does not only include money, beds or
professional personnel. It must also involve self-
and community assistance. The same disclaimers
apply. Circumstances will dictate which resources
are available, needed and utilised by whom and why.

To a broad conception of health must, therefore, be added broad ones of care and resources. In some ways, this approach is cogently summarised by McKeown (1979, 79):

> The appraisal of influences on health in the past suggests that we owe the improvement not to what happens when we are ill, but to the fact that we do not so often become ill; and we remain well not because of specific measures such as vaccination and immunisation, but because we enjoy a higher standard of nutrition and live in a healthier environment ... In the light of these conclusions the requirements for health can be simply stated. Those fortunate enough to be both free of significant congenital disease or disability will remain well if three basic needs are met; they must be adequately fed, they must be protected from a wide range of hazards in the environment, and they must not depart radically from the pattern of personal behaviour under which man evolved, for example, by smoking, overeating or sedentary living.

Health concerns, therefore, diet, environment and self-care. We would add medical care to the list of necessary requirements. That type of care is a significant element in achieving happy personal and social development, the bases of good health (see Wilson, 1975). While the perfect health, implied in its definition as a residual category, is an impossibility, good health is not. Health state derives from the dynamic interplay of personal attributes and life-circumstances, integral parts of which are types of care and levels of resources.

Care

Just as we saw that health was given a specific meaning, we may note that care is also viewed partially. Health really specified lack of treatable illness; care means treatment or cure. In more general terms, care is taken to mean some kind of intervention by certain categories of professional workers. While not wishing to underemphasise the important and significance of care as intervention, as with health, we wish to remove the definitional assumptions to explore the nature of care more fully in line with our commitment to a broadly conceived health and health care.

In everyday speech, care means to be concerned or interested. Like all feelings or attitudes, it is directed at a particular object or objects. In the field of health, care relates to people, not just others but also ourselves. We may say then that care is self-directed and other-directed. It should be noted that we say 'and'; care is not either self- or other-directed. It is both. We recognise that the dualism of altruism and individualism or egoism is a false one. It is, however, not our purpose to become embroiled in the debates over whether altruism is innate or how its existence can be explained when it seems to militate against personal survival and well-being. But sociobiologists such as Wilson (1975) do demonstrate, perhaps despite themselves, that acts and feelings of altruism are contextual in terms of both the specific circumstances and social mores and values. As Midgley (1979) notes altruism is a regard for others as a principle of action. And as of course all principles of action are normative, altruism must be contextual. Part of the context is conventionally defined in the sense that guidelines for action are built into particular roles and role performances (see Emmett, 1966). A role description - father, sister, doctor - in itself implies the existence of certain rights and obligations connected with that role. Altruism, care and compassion may, therefore, be seen as being embedded in social relationships as structured by roles. In this sense, care and caring become dimensions of the moral fabric of a society, enhanced or retarded by the nature of socialisation and social development. The nature of care depends on how people are taught to think about and act towards other people.

It may be thought that such a comment does not tell us a great deal about the nature of care. On the contrary, it tells us, in general terms, everything. 'Care' is a social construction and like all such constructions once established it shapes the lives of those who 'built' it. People do not get the care that they deserve but the care that they themselves help create. The framework of care is the normative framework of society, part of which, as Plant (1970) avers, is the presence of social and role morality. Social morality is not detached and amorphous but discrete, related to the performance of roles in society. Plant is concerned with the relationship of social workers to their tasks and clients but his comments have more general applicability. He notes that the main issue concerns the

relationship between socially accepted standards
which are attached to certain performances and the
principles, ideas, beliefs and decisions of individuals who actually perform these roles. This issue
seems central to any caring or to care itself. The
multiplicity of roles and the divergences in moral
stances in the modern world - the 'plurality of
life-worlds' as Berger et al (1974) put it - means
that there are few agreed-upon ways of thinking and
acting. Indeed, there may be no universal guidelines, even respect for human life, as Turnbull
(1973) found in his study of the impoverished moun-
tain people of East Africa, the Ik, whose dire
circumstances meant that the old and very young were
not only denied sustenance but also robbed of it.
These schisms in how we ought to act is, however,
particularly important in societies, such as
Britain, dominated by secondary, contractual social
relationships. The problem and the often-taken
solution is excellently summarised by Emmet (1966,
132)

> ... a morality simply of direct I-Thou
> relationships cannot take account of the
> host of indirect personal relations in
> which we stand to people, nor the imper-
> sonal element which arises even in a per-
> sonalist morality, and a fortiori in the
> morality of official institutional rela-
> tions. A purely personalist morality since
> it cannot come to terms with these must
> abandon them to anarchy (which is
> unrealistic) or to external regulation
> which may be all too realistic.

While we do not accept that abandonment as part
of 'modern morality' is unrealistic (out of sight,
out of mind), we recognise the strength of Emmet's
comment concerning external regulation. In this
sense, care is doubly created, first by the moral
order of society and secondly by the formalisation
of the caring relationship in certain professions.
With formalisation, care may be seen as intervention. The formalised caring relationship has been
explored in the health field by sociologists (see,
for example, Freidson 1970; Mechanic, 1978). It
has also been examined by Downie and Telfer (1980)
who note that the 'bond' which constitutes the
relationship in the social or medical services
consists of formal legal and administrative rules
and a vaguer set of rules and expectations which
doctors and patients, social workers and clients
have of each other. The professional carers often

refer to his second set of rules as their 'ethics' and it ideally involves impartiality, objectivity, non-judgementalism and compassion. But as Downie and Telfer argue in their discussion, these 'ethics' are not absolutes and may be individually inter- preted. Personal characteristics, situational circumstances and societal setting will affect the nature of these informal rules. Thus, although the professions would like to regard these values as being 'above' individual practitioners and the social order, they are individualised and (of most interest to us) are pre-eminently social.

In a society like Britain, the values of impar- tiality, objectivity and non-jugementalism (ie those of professionalism) tend to be emphasised at the expense of compassion, which must have an ideational as well as practical basis (see Robson, 1976). Material gains rather than the moral basis of social life dominate, a point eloquently made by Seabrook (1982) as he sees the modern unemployed not only losing their material well-being but also not having the support, compassion and shared spirit of the traditional working class to fall back on. In a way, religion has been replaced by materialism as the source of guidelines for action. The decline of the former has not been matched by the moral rise of its rivals - socialism, communism or pacifism. To argue that the basis of care have declined is not the same as suggesting that they no longer exist or that altruistic actions do not exist. Titmuss (1973) exemplifies the existence of altruism in modern Britain in, what he terms, the gift relation- ship, namely the donation of blood to others for no payment. Collard (1978) is, however, surely correct when he argues that the opportunity cost of giving blood is low. In Titmuss' scheme, the technical parameters of the service - the 'external regula- tion' - do not loom large. But in this instance care is other-directed and motivated by feelings of altruism, reciprocity and duty. We must be wary of applying this definition of care and these values to other aspects of provision or intervention, although we may note that such definitions and values also appear to underpin the overwhelming public response to community disasters at home and abroad. It may indeed be the scale of the misfortune that allows such values to be expressed. In such circumstances, people are an aggregate. They are not individuals whom we must confront on an individual basis. These values are other-directed but in an amorphous way. Many problems and illnesses are, of course, everyday

occurrences in individual lives. In such conditions, we are unsure of our role(s) because of the
lack of certainty over guidelines for action. We,
therefore, as individuals, withdraw, leaving the
field for professional carers (those whom we pay to
care) or to family and 'community'. Family and community care will be treated as resources below.

Some of the problems emanating from the relationship between 'carer' and 'cared for' have been
treated by Pinker (1971) who regards a welfare
system as one of unequal exchange. Indeed, we would
argue that all care relations (actual and potential)
should be so viewed. Caring is based on a reciprocity yet any form of stratification or inequality
destroys the egalitarian basis of reciprocity. The
wealthy may give or provide care but the outcome is
not straightforward. The less wealthy receive but
the act of giving or caring, whether motivated by
self-interests or altruism, enhances the prestige of
the giver or carer and may stigmatise the recipient
(see also Lenski, 1966). Maus (1954) notes the
presence of such relations in modern charity.
Pinker (1971) extends the argument to modern care
systems seeing them as generators of stigma and
subordination. Indeed as most people in modern
societies are denied access to effective positions
of power, they cannot pursue such prestige.
Instead, as the providers of 'care' through
taxation, they impose stigma. Such stigmatisation
is enhanced by the vigorous means-testing of
benefits. To apply is in itself stigmatising and
this has the intention of deflecting the resentment
of the subsidised by the lowest wage-earners and
tax-payers (see Jordan, 1974). Further, it is
interesting to note that the spatial relations
between giver and receiver are significant in the
process of stigmatisation (see Pinker, 1979). Close
proximity not only increases the propensity to care
but also the capacity to impose sanctions. Indeed,
much social work intervention concerns the reduction
of the 'distance' between caseworker and client
through investigations of personal circumstances.
To know may well mean being able to control the type
and quantity of care provided and to define the
conditions under which care will be provided.

The degree of stigmatisation is not, however,
fixed. Much depends on the values that permeate
society and on the interactions between different
social constructions of reality. Complex industrial
societies like Britain do not have one world-view or
one model of care. The definitions and experiences

of different groups interpenetrate and confront the
dominant meaning-system. But as Williams (1977)
points out, this hegemony is itself full of schisms
and contradictions. It maintains its currency not
by coercion or manipulation but by suggesting that
certain ways of thinking and doing are 'natural' and
'correct'. Even then in democratic societies this
meaning-system can be 'captured' and changed to suit
the interests of an ascendant social grouping. (We
accept, however, that no grouping has in Britain
challenged the fundamental interests of the capitalist system: different groupings have different
views of how best to advance capitalist growth and
accumulation, the route for those that have held
political power to universal prosperity.) These
changes can be found in different views on the
nature of care. A brief consideration of these
views will also enable us to confront other dimensions of 'care', namely the degree of intervention
necessary in other-directed care, the public as
opposed to private provision of forms of formal care
and the relative significance of self-care and
individual responsibility in enhancing health.
   As ideal-types, we may point two models of care:
the residual and the institutional:
       The first holds that social welfare insti-
       tutions should come into play when the
       normal structure of supply, the family and
       the market break down. The second, in
       contrast, sees the welfare services as
       normal 'first-line' functions of modern
       industrial society. (The models) represent
       a compromise between the values of economic
       individualism and free enterprise, on the
       one hand, and security, equality and
       humanitarianism on the other (Wilensky and
       Lebeaux, 1965, 138-9)
While the first part of the quotation refers to
formal welfare systems, the second part demonstrates
that these (and all care) are underpinned by certain
strategic values. 'Institutional' care refers to
the post-1945 consensus on care and welfare. Every
citizen has the right to assistance from the state
if s/he is unemployed, old, disabled or sick. Such
care emphasises collective public provision of
services and is based on an interventionist strategy, providing an unconditional 'national minimum'
for all. In other words, universalism rather than
selectivity is the basis for the allocation of care.
In practice, this ideal-type has tended to break
down with means-testing, selective intervention and

17

cash-limiting being introduced. Residual care sees
state intervention as a safety net for the poorest.
Self-reliance, individual responsibility and the
concomitant use of private health care plans are
seen as the eternal verities. It is a model based
on the poor laws of the nineteenth century which saw
universal state provision as the way to national
economic and spiritual decline. Its model of care
was, however, well summed up in 1973:

> ... the only real lasting help we can give
> to the poor is helping them to help them-
> selves; to do the opposite, to create more
> dependence, is to destroy them morally,
> while throwing an unfair burden on
> society.
> (And) Parents are being divested of their
> duty to provide for their family economi-
> cally, of their responsibility for educa-
> tion, health, upbringing, morality, advice
> and guidance, of saving for their old age,
> for housing. When you take responsibility
> away from people, you make them irrespon-
> sible. Hand in hand, with this, you break
> down traditional morals. (Sir Keith
> Joseph, quoted in Jordan, 1976, 137, 138).

Both these models claim to be caring and compassionate, the institutional in the commonsense way of
being immediately concerned about, the residual
being compassionate at one remove: 'we care, therefore we do not intervene'. While we abhor such a
stance, we may note in passing that it does demonstrate the necessity of seeing care as a social
relationship rather than a just and necessary intervention. These two views of care will be met again
in our examination of allocational mechanisms in
Chapter Two. We must note, with Jordan (1976), that
both views see care as being concerned with weakness, the institutional with trying to protect
individuals from certain inadequacies and weaknes-
ses, the residual regarding care itself as weakness
except for the weakest in society. Both views have
the capacity to stigmatise the 'cared for', although
we would assert that it is a far stronger tendency
in the residual model.

But just as receiving care is not necessarily
stigmatising nor giving necessarily prestigious,
care itself must not be seen as self-directed or
other-directed. Nor is its provision individualist
or collectivist, nor based on selectivity or universalism. And of course in practice, the dichotomy of
institutional or residual itself breaks down. Care

can be and is all of these things. It is a social
construction, embedded in social relations and
structured by the moral fabric(s) of the social
order. As a social construction it will be contextual and contradictory. Its broad nature will be
established by social parameters but its specificity
will depend on circumstances, individual, situational and societal. Any analysis of care as cure
or as intervention must recognise the contextual,
relational nature of caring. Just as our conventional view of 'health care' assumes a particular
definition of health so too does it posit care in a
partial way. This broad-based view of care must be
continually recognised in our analyses and criti-
cisms of 'health care policy.' And further while
'care' implies a social relation, it also possesses
a 'resource' dimension. In broad terms, 'care' may
be seen as orientations to action, while 'resources'
may be taken to mean the ways and methods of acting:
both orientation and method being in relation to the
enhancement of health as well-being.

Resources

In conventional terms, 'resources' means money and
the conversion of these financial resources into bed
numbers, drugs or personnel. Resources represent
the input into the formal care system, although as
we shall see, a more broad conception of resources
still allows their treatment as inputs or the means
to health. If, however, we simply consider financial resources, we may see that it is possible to
ask some pertinent questions, especially from a
geographical perspective. It is with respect to
resources that we may talk of a geography of the
national health service. Financial resources are
converted into health-related goods and services
(treatments) and then allocated between groups,
sectors and territories. Indeed an examination of
financial resources leads immediately to questions
of resource distribution. Not only do such
resources, because of their common denominator as
cash, result in an implicitly integrated approach to
public policy through public expenditure plans (see
Chapter Five), but also raise the issues of who gets
what, who should get what and how big is the 'what'
that is available and received. We leave the
question of how resources are allocated to Chapter
Two, and we subsume consideration of where under
'who', which is ungrammatically defined as a client
group, service, sector or region.

Questions of distribution become important in situations of resource limitation and scarcity. All governments are concerned with rising costs of health expenditure and the rising curve of financial resource provision is likely to flatten in the future (Office of health Economics, 1961). Indeed, Maxwell (1981) shows that even for the period 1975-7 only Sweden significantly increased its proportion of GNP devoted to health expenditure. In most capitalist nations, the proportion remained unchanged or declined as in Britain (-0.3 per cent). But total health budgets have not declined in reality (see Klein, 1983), despite the perception of decline. That perception may emanate from the centrality given to cost-containment policies by governments. It may therefore be a case of what Wildavsky called 'doing better and feeling worse', of resources increasing but few perceived benefits accruing. Indeed although the NHS has remained fairly sheltered from the rigours of the British economic climate because of its continuing public popularity and support, cost containment is the primary watchword and target allocations are now provided for services to particular groups in the regions and districts. While some, like Klein (1981), have been charitable suggesting that this 'delegation' of responsibility for resource distribution is the most sensible course of action in times of economic uncertainty, it also means that any unpopularity for the decisions is also devolved. In times of economic stringency, distributional questions become increasingly acrimonious and in the British context that acrimony occurs not only at the national level, but also at the regional and local levels.

This devolution of acrimonious debate is part of the strategy of governments "applying the brakes to their health care budgets as firmly as they judge politically possible" (Maxwell, 1981, 47). It is in the phrase 'politically possible' that we may see that resource limitation and scarity are in part at least politically and socially determined. It must of course be recognised that the resources of the earth are finite (although capable of technical extension) as are those of any one nation (although capable of extension by increases in productivity and controlled borrowing). But scarcity is itself a social construction. We accept that at some point there exists a state that may be called absolute or ecological scarcity, defined in terms of a finite material base, finite space and finite labour-power

(see Brookfield, 1975). But scarcity is also insti-
tutionally produced with particular institutions and
sets of relations determining who gets what and
rationing the allocation of resources to particular
activities, groups and territories. The social
scarcity (or limitation) of financial resources to
the health care sector is a matter of broad economic, political and fiscal policy. Within the health
field, it is a matter of national and regional
debate, part of which, in a system of cost-containment and rationing, must entail the establishment of
'priorities'.

This establishment is of course part of the
wider distributional question. Various documents
have been produced on priorities for health care
resources (DHSS, 1976b; 1977b; 1981) and elements
of these policies are appraised in Chapter Five.
Clarity at the centre often clouds at the periphery
and despite the recent requirement that RHA and DHA
chairmen have to report on progress, the lack of
target allocations for specific services means that
policy is difficult to implement. National policies
are accepted but the regional and local response is
slow, uncoordinated and often non-existent (see
Butts, et al, 1981), although the appointment of
general managers (similar to those operating in
commercial circumstances) in the NHS is a market-
oriented strategy to improve co-ordination and
decision-making abilities. The role of regional and
district authorities had been mainly reactive,
concerned with local implementation and problems
rather than grand strategy and influenced greatly by
the wishes and demands of their professionals and
managers (Klein, 1982).

Prioritising in times of cost containment
creates additional acrimony. Financial resources
and their expenditure are assessed in the greatest
detail, and the requirement to shift resources takes
place against this backdrop. Reallocation means
that there are often real losers, as additional
funds are seldom made available in the budget for
those defined as 'needy' and requiring growth and
development. Acute general and maternity services
must lose relative to the priority dependency groups
such as services for the elderly, the mentally ill
and the mentally handicapped. The 'rich' regions
such as those in the London area lose relative to
'poorer' ones such as Northern and Trent. The
issues and problems of the formulae for allocating
resources to the regions have received various and
critical treatment in terms of both technical and

equity criteria (see Maynard & Ludbrook, 1980;
Jones and Prowle, 1982; Eyles and Woods, 1983).
The largest increases in financial resources are
given to those regions which are furthest below
their targets. Smaller increases should accrue to
the 'richer' regions but in recent times, this has
resulted, particularly at the district level, in
real losses in resource allocation. These districts
have great difficulty then in allocating resources
between particular sectors and services. While the
regional resource allocation strategy has resulted
in a narrowing of regional differentials so that the
gap between worst and best has narrowed from 22 per
cent in 1979-80 to 14 per cnt in 1984-5 (DHSS,
1984), we are unsure what this quite means in terms
of health care provision (see Figure 1.1). Of
course we recognise that the health status of those
in the Northern and Trent regions should be improved
(although the nature of the relationship between
health status and health expenditure is unclear),
but note that greater resource equity at one scale
may lead to greater inequality at another scale.
The formula takes account of such factors as age and
sex structure, SMRs, fertility rates and marital
status but does not include a class or deprivation
measure. Even if the actual constitution and effect
of such a measure remain problematic (see Woods,
1982), a commonsense view must surely aver that the
removal and reduction of financial resources from
inner city districts in 'rich' regions are nonsensi-
cal acts. The pursuit of equity is a difficult
task; its pursuit in circumstances of cost containment and resource limitation is virtually impossible. These apparently contradictory aims may be
one of the reasons for the 'doing better, feeling
worse' syndrome. But not all parts of the service
are doing better in terms of resources. And 'feeling worse' allows us to raise other issues of
resource distribution because it is not only the
quantity of resources that is important but also
their quality.
     The quality of resources relates to the pheno-
mena into which financial resource are converted.
It is in a consideration of quality that 'care' and
'resources' become conjoined as much depends on the
effectiveness of treatment and the morale and commitment of personnel. Effectiveness is extremely
difficult to measure and as Cochrane (1972) notes
few procedures in health care have been rigorously
evaluated and in any event most evaluations rely on
habit, custom and tradition rather than rationality.

Figure 1.1 Resource allocations 1977/8 - 1984/5

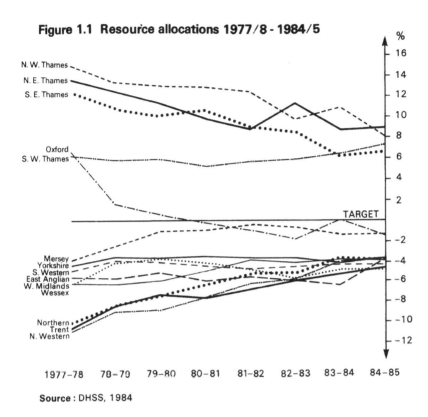

Source : DHSS, 1984

We may also note that if we judge quality merely
through effectiveness which may itself be measured
by number of beds or average length of stay in
hospital or number of GP consultations, we become
embroiled with efficiency criteria too. Necessary
quantifiable assessments can then be used as the
basis for making financial judgements on which
service to fund. It is, of course, difficult to see
how doctor-patient and social worker-client relationships can become elements of resource distribution policy. Such relationships do however form an
important dimension of the quality of resources and
care. The personnel in the health field practise
their skills as mediated by their values. Their
practice is in part their compassion for and commitment to care. We may note that these values have
been increasingly formalised not only by professional training but also because of the politicisa-
tion of health issues and the unionisation of health
personnel. 'Care' becomes rationalised as its providers - the resource - see their work as little
different from any other form of employment. Further, the circumstances in which the carers provide
their labour means that they have to ration their
time and involvement with clients by informal procedures. Parker (1975) identifies four mechanisms.
These entail deterrence (discouraging prospective
patients from using and demanding services), deflection (referring people to other workers or the
community), dilution (reducing standards or spreading services more thinly than the ideal), and delay
(rationing time through appointment systems and
waiting lists, necessary features in a demand-led
system which have though been greatly extended for
reasons of financial exigency). So cost-containment
entails rationing the quality as well as the quantity care.
    Considerations of personnel extend 'resources'
to include their quality as well as quantity. We
may use the notion of 'deflection' to extend it
further especially if we see it as a form of
resource substitution. As Wilkinson (1973) notes in
his study of British and American economic development, conditions of scarcity may lead to resource
substitution. If the formal sector resources are
not available or are limited or delayed, others may
take their place if the need is great enough. These
resources are primarily those of self-treatment and
family and community care because "societies adopt
the new resources and methods which come most easily
to hand in the context of their established cultural

system" (Wilkinson, 1973, 101). This view of
resources must consider their spatial relations.
First, however, resources must be broadly seen as
social and technical appraisals (see Spoehr, 1956).
Harvey (1973) suggests that this means that individuals must possess the various cognitive skills and
technological equipment to make use of 'resources'
and value systems to motivate them to use them.
These appraisals affect an individual's command over
resources, which is itself the basis of real as
opposed to monetary income. This command is a
function of locational accessibility and proximity
in two senses. First, individuals possess a location relative to environmental goods and bads which
may enhance or endanger health and well-being.
Secondly, proximity to like or particular kinds of
people may embed an individual in an informal care
network, although in this sense proximity must
include social as well as spatial considerations.
Bell (1968), for example, noted that despite their
geographial separation, the members of middle class
families still formed networks of mutual aid and
care. They provided one another with resources
despite their geographical mobility and because of
their social proximity. While it is difficult
(though not impossible) to quantify the value of
these personal networks, they are significant elements in any 'health care system' and of any
schedule of resources. Indeed, they form one of
five dimensions of resources (see Townsend, 1979).
With home production (ie smallholding or garden) and
gifts, 'personal supporting services' form part of,
what Townsend calls, 'private income in kind', an
integral part of an individual's resource system,
which may also contain cash income, capital assets,
employment benefits in kind and public social services in kind.
    Personal supporting services, including self-
treatment, are extremely important. In fact, Last
(1963) used the term 'illness iceberg' to denote
that most symptoms do not lead to a medical consultation. Further, lay people (usually relatives) are
used as a sounding board in about three quarters of
those symptoms that lead to a consultation with a
doctor (Suchman, 1964; Scambler et al, 1981).
Social networks are, therefore, seen as important
support resources (see Cobb, 1976; Pilisuk and
Froland, 1978). Much of the research has, however,
focused on the roles played by these networks in the
process of illness behaviour, and therefore whether
different types of network make individuals more or

less likely to consult professional carers.
McKinley (1973) for example, examined the influence
of family and friendship networks on the use of
maternity services in Aberdeen, noting that under-
utilisers appeared to rely more on readily available
relatives and friends acting as one large inter-
locking network (see also McKinley, 1981). Family
and friend networks operate in different ways, with
large family networks tending to inhibit medical
consultation and large friendship ones precipitating
it (Salloway and Dillon, 1973). With psychiatric
problems, Horwitz (1978) found that friends wanted
to connect the individual sufferer with the professional network while family tried to maintain
him/her within the informal sector. The undoubted
importance of such networks as resources is demonstrated by Totman (1979) who points out that the
most vulnerable individuals in society are the
marginal, the isolated, and those with changing or
ambiguous role structures and with inadequate social
support.

It is perhaps the 'family', although it can be
variously shaped, that is still the main locale of
our supporting services and therefore of our caring
commitment. Such resources and supports are of
course variable as all families construct to some
extent their own reality. But 'family care' is
significant for living and coping with problems.
Bayley (1973) in a study of care for the mentally
handicapped notes the nature of mutual aid amongst
kin, the reluctance, based on the fear of stigmatisation, to admit to needing outside help, and the
reluctance to allow public intrusion on private
matters. Family care is not unproblematic. Nor is
'community care', which implies the assistance of
friends and neighbours as well as family to provide
care and enhance health as well-being. This assistance has been and is forthcoming, particularly in
traditional working class and occupational communities. As Parker (1975, 27) puts it "... areas where
families and individuals are bound together by
kinship, common residence and often common occupation, develop elaborate 'private' welfare systems
based on mutual independence and resistance to
public services". Such 'informal welfare' is often
informed by exclusions, sanctions, conditions and
problems as well as care (see Allan, 1983). Family
care often places great strain on women, because of
the role of men as chief wage-earners and because
femininity is equated with caring. This strain is
particularly intense where long-term commitments, as

with the elderly infirm, are required. Community
care may be sanctoned in that attachment to communal
ways of thinking and acting may be a condition of
the availability of care. Further, the capacity of
families and communities to provide care may well be
limited. This point is not recognised in the formalisation of community care, now seen by providers
and politicians as a 'cost-effective' (ie financially cost-containing) way of providing care.
Community care is being co-opted and as Graycar
(1983, 382) notes "rhetoric and reality in the field
of informal tending are quite distinct. Cost-
cutting politicians exhort us to return to a golden
age where families provided a greater amount of care
than is assumed they do today. The reality is that
there are severe limits on family capacity to do
so." It is a tragedy that this resource - informal
welfare - will be manipulated in this restructuring
of care. Indeed, the coincidence of conservative
(cost-containment) and liberal (lay participation in
care) views demonstrate the fact that the resource
is being seen and used in particular, partial ways.
It would be a double tragedy if the present pre-emi-
nence of the conservative view led to socialist
arguments against the concept of community care in
its entirety. Such arguments, like those which saw
the welfare state as an irrelevance or vehicle to
enhance capitalist development, may lead to the
demise of this significant arena of care. As an
East Ender said "I take more notice of my mum than I
do the welfare" (quoted in Offer, 1984, 554). We
shall examine the partial definition of community
care as an alternative cost-effective integrated
care in Chapter Five. But our purpose here has been
to extend the notion of 'resources' not to replace
the conventional primacy of financial resources with
one consisting of self-help and localised mutual
assistance. Both are part of our conception of
health care resources.

Conclusion

It has been our purpose in this opening chapter to
explore the three concepts that we feel have a
significant bearing on the 'national health', these
being health, care and resources. It will have been
noted that we advocate broad conceptions of all
three phenomena, this advocacy being based on our
premise that good health constitutes a good quality
of life which can be enhanced by the availability
and use of a resource system, the nature of which is

itself shaped by particular orientations to action,
particularly the caring relationship. Health was
therefore seen as well-being. There are indeed
close parallels in terms of concept of interest with
the welfare approach in human geography (see Smith,
1977 and Chapter Six). Starting-points and means of
identifying well-being do, however, differ. Our
starting-point is 'health' rather than the isolation
of individual variables and ours means the analyses
of the bases of health and care delivery rather than
the statistical description of states of territorial
well-being. Important as this work is, we may
comment that we start where the welfare approach
ends. And our analysis proceeds by discussing care
and resources, the orientation from and means by
which particular states of health as well-being
derive. Care was seen as a social relation while
resources were seen not only as money but also as
technical and social appraisals. It must be recognised that we do not reject the conventional ways of
examining health, care, and resources. These ways
are limited and as such are contained within our
broader conceptions. So health as the lack of
illness is a dimension of health as well-being,
while care as cure or intervention is one element of
care as a social and moral evaluation and relation,
and financial resources and their conversions are
implicated in resources as appraisals and actions.
Further, we do not see health, care and resources as
independent phenomena. We have noted in the fore-
going discussion the interrelations of care and
resources: both pertain to health as well.

Any geography is concerned with pattern and
distribution as percursors of process and explanation. Ours is no exception. Indeed, of that descriptive question of brilliant simplicity - who gets
what, where and how - it can so far be said that we
have tackled the 'what' in this exploration of
health care resources. To advance our discussion of
the geography of the national health we must now
confront in more explicit detail, the who, where and
how. We shall do this by examining the bases and
nature of health, illness and health care in
Britain. Chapter Two will look at the 'how' with
respect to the allocational bases of health and
health care. It focuses on the mechanisms used to
allocate health care and the notions of justice and
need on which these mechanisms are themselves based.
Chapter Three addresses the 'who' with respect to the
social bases of health and illness. The 'where' is
addressed throughout these chapters (as a vital

dimension of who (people in space) and how (allocation across space). We turn to the theme of
'integrated health' in Chapters Four and Five.

Chapter Two

ALLOCATIONAL BASES OF HEALTH CARE

Having looked at the 'what' part of who gets what,
where and how in Chapter One, we now concentrate on
'how', i.e. on what bases are health care resources
distributed among different groups and territories?
We shall of necessity emphasise the allocation of
financial resources because personal support services depend greatly on the nature of family and
group life. The support that families and kin and
friendship networks can provide is constrained by
the financial resources at their command, primarily
their income. We may note, however, that other
'resources' such as knowledge, reciprocity of services, and emotional support and commitment may be
available and used. This non-financial assistance
is an important dimension in the 'making do'
strategies so ably described by Hoggart (1957) and
seen as part of the 'community of the oppressed' by
Williams (1971). We as yet know little about how
these 'resource systems' operate to enhance the
well-being (or rather militate against ill-being) of
particular groups and networks. Anthropological
investigations of small-scale societies show the
intermeshing of financial resources (wealth) and
other forms of assistance as central in shaping
social life and determining quality of life. Such
examinations of industrial societies are beginning
to demonstrate similar features whether of a neighbourhood joining together to try to deflect and
delay what are seen as deleterious 'outside' forces
(see Susser, 1983) or of households simply living
their lives as best they can (see Wallman, 1984).
We recognise that the resources found in such communities are not simply 'there'. Nor are they
provided to all individuals on a similar basis or to
a similar degree. In other words, these support
services are allocated in terms of particular

'principles' which orient and shape group life.
Such principles - e.g. power and sanction, personality and predisposition, intersubjectivity and
interaction - are beyond the scope of this book,
involving a social psychological dimension to provision as well as a social structural one. There is
what may be broadly defined as an anthropology of
care and provision, which we intend to address in
general terms leaving its specificity to later
investigation because it entails detailed comparative analysis. But the general examination means
that we are able to look beyond the question 'how'
to 'why'. Thus why are health care resources allocated to different groups and territories? And why
does this allocation occur in the ways that it does?
Such questions involve not only a consideration of
allocation but also need and justice.

Need

The idea of need is fundamental to understanding
social policy, of which the allocation of health
care is a part. As Bradshaw (1972, 640) notes "the
concept of social need is inherent in the idea of
social service. The history of the social services
is the history of the recognition of social needs
and the organisation of society to meet them."
Needs are in fact part of being human. Their centrality is assessed as follows:
> the regulation of human behaviour in the
> pursuit of needs and wants is the prime
> source of social relationships, political
> institutions and modes of production. It
> is the source of our codes of morality, of
> the laws derived from them, and of the
> various other ways in which we try to
> resolve conflict. The individual human
> being will presumably view his or her life
> quality in terms of the extent to which
> perceived needs and wants are satisfied
> (Smith, 1977, 27).

It may be noted that this discussion correctly
indicates that needs are societally determined and,
problematically, that they are related to wants. In
everyday speech, we tend to assume a continuum of
requirements and aspirations from basic needs to
wants of a fantasising kind. We should not, however, attempt to make a distinction between physiological and other needs. As Plant et al (1980)
show, even establishing physical survival as the
basic human need leads to at least two ends rather

31

than one clear-cut conclusion. It may lead us to
help one another to survive or to view our obligation as being simply not to kill. And this latter
view may be taken further to see our duties not as
positive and other-directed but as those imposed on
ourselves by voluntary contract (see Nozick, 1974).
This contract view of need may be said to inform in
part market principles of resource allocation (see
below). The example of physical survival thus
indicates that obligation is a social construction,
so too are 'physiological requirements'.

We cannot of course survive without food, water,
air, warmth and sexual activity, these being in
Leiss' (1976) terms, species needs. But there is
great diversity in how different groups respond to
these physiological requirements. Even with food,
there are sets of culturally determined practices
that select, prepare and avoid the substances which
can serve as the bases of nourishment (see Levi-
Strauss, 1970). All human needs are routed indirectly towards sources of satisfaction by culture,
blurring any distinction between basic and derived
needs. As Lee (1959, 72) avers, "it is value, not a
series of needs, which is at the basis of human
behaviour." This view sees needs themselves as
derivative, an analysis in part taken on board by
attempts to establish hierarchies of needs. The
most influential ranking of need is that of Maslow
(1943; 1954), who arranged needs from 'lower'to
'higher' and from 'more immediate' to 'less immediate.' Higher needs emerge serially after the more
basic ones have been adequately met. There are
three basic or 'deficiency' needs, these being
survival (the physiological struggle to sustain life
by obtaining food, clothing, shelter and so on),
safety (protection from physical danger, the provision of order and the predictability and dependa-
bility of the environment), and love, affection and
belongingness (the need for affection, meaningful
social relationships, group harmony). We may note
that health as well-being is denoted by all three of
these deficiency needs, involving as it does body,
mind and spirit. The other two needs - self-esteem
(the need for recognition, status and prestige) and
self-actualisation (the desire to live up to one's
potentialities and capabilities) - are required for
a full personality to develop. Maslow's scheme is,
however, societally specific, reflecting the organisation of life in a technologically advanced
society, where there is a high degree of specialisation among functions and activities (see Leiss,

1978). This view is given implicit support by Dahl and Lindblom (1963) who suggest that Western societies emphasise existence or survival, psychological gratification (through food, sex, sleep and comfort), love and affection, respect, self-respect, power and control, skill, enlightenment, prestige, aesthetic satisfaction, excitement and novelty.

Needs and their satisfactions are in many ways social constructions. Need is a historically and culturally relative concept. This relativity makes the distinction between needs and want extremely difficult to draw. It has been argued, for example, that wants are states of mind whereas needs imply a failure to attain some agreed standard (see Benn and Peters, 1959; Braybrooke, 1968). But where does the state of mind come from and against what is a particular state deemed satisfactory or unsatisfactory? This is not just a general moral and societal issue but one directly relevant to medical care itself. Through the medicalisation of social life and the inclusion of psychological health within the preserve of medicine, this form of care must deal with states of mind as wants and with 'higher' order needs. Can a patient 'wanting' 'peace of mind' be equated with the 'necessary' treatment of a person diagnosed as a schizophrenic? Further, technological and biochemical advances in medical treatment mean that more conditions can be treated. If the satisfaction of a 'problem' becomes attainable, does the condition become a need? Thus, not only is 'need' inextricably linked with its satisfaction but also with the supply of the means of this satisfaction - a point to be explored below. Needs and wants are, therefore, related in complex ways. We do not subscribe to the view that needs cannot be satisfied because of a revolution of rising expectations i.e. increasing wants. This is simply the economic case of blaming the victim (or the individual). Galbraith (1974, 174) is surely correct when he says that "as affluence increases, goods become increasingly dispensable or even frivolous." Further, people judge themselves and their positions in society against that which others have and society can provide (see Runciman, 1966; Townsend, 1979). Consumer capitalism not only requires that its goods and services be wanted and bought but also defines individual identity and actualisation in terms of consumption. Wanting thus becomes 'natural' and all 'systems of needs' become relative and not accountable as such by physical necessity.

Allocational Bases of Health Care

The relative and societal dimensions of 'need'
mean that most needs and wants (as defined above)
will be felt as real. We accept them as real and
reject the notion of false needs: "those which are
superimposed on the individual by particular social
interests in this repression" (Marcuse, 1964, 5).
Cause may well be conflated with effect. The effect
of pursuing certain needs may well enhance the
capacity of a social order to reproduce itself.
This is not the same as saying that interests in a
particular order deliberately and consciously incul-
cate the population with ways of doing that serve
themselves. The reproduction of labour-power and
existent capitalist relations are enhanced by the
provision of state health care resources (see Gough,
1979) but this effect cannot explain why the non-
productive population are allowed access to any or
similar resources. The rejection of 'false needs',
also seen as 'artificial wants', means that the
distinction between needs as objective requirements
and wants as subjective states breaks down. This is
important for policy as well as conceptual reasons.
It is worth quoting Leiss (1976, 72-3) at some
length on this point:

What is detrimental about the attempted
demarcation between needs and wants is that
it encourages us to regard the sphere of
needs largely as a quantitative problem:
each person needs a certain amount of
nutrients, shelter, space, and social services. The practical outcome of this
statement of basic needs is reflected in
some of the social policies of the existing
welfare state: bulk foodstuffs for the
poor, the drab uniformity of public housing
projects, and the stereotyped responses of
bureaucracies. The qualitative aspects of
needs are suppressed in these policies,
just as the qualitative aspects of needs
are suppressed for society's more fortunate
members in the quantitative expansion of
the realm of commodities.

The satisfaction of need is seen in care as treatment or intervention (Chapter One) and the success
of intervention is seen in the number of interventions. In emphasising quantitative satisfaction,
care becomes a matter of supply and need is manifested in demand. The societal definition of need,
therefore, becomes a double construction, in terms
of the social basis already discussed of what a
society can provide and achieve with respect to

existing parameters. What can be achieved rather
than that which is achievable is the focus for
attending to need. This may well be a commonsensi-
cal view in terms of available financial resources
but it is a view which constrains the definition of
care, the greater attainment of health as well-being
(this being recognised as an ever-changing and
therefore unattainable goal) and the constitution of
the 'good society'. In concrete terms, it blinkers
policy and in the context of limited financial
resources means that what are needs can be redefined
as desirable, i.e. as wants, which will be satisfied
when funds become available. Thus the quantitative
definition of needs and wants recognises in practice
little as being of intrinsic importance. Financial
exigency eats away at wants and then at needs. And
it is not simply a matter of replacing public routes
to need satisfaction with private ones. An individual cannot enforce his/her own safety at work,
cannot be sure s/he is eating unadulterated food, or
determine emission levels of industrial waste
materials into the atmosphere or water. We may have
chosen issues of an uncontestable nature as illust-
rations, although all these programmes are under-
funded in Britain.

What is intrinsically important may in itself
not be contestable. Indeed, we follow Miller's
(1976) definition of need which emphasises felt need
and over which there may be little disagreement. He
defines need in relation to harm, which "for any
given individual is whatever interferes directly or
indirectly with the activities essential to his plan
of life and correspondingly, his needs must be
understood to comprise whatever is necessary to
allow these activities to be carried out" (Miller,
1976, 134). This is similar to Wollheim's (1976)
ailing and Peters' (1958) view of need as a normative conception, prescribing a set of standard goals
and functioning as a diagnostic term with remedial
implications. What are of course contestable and
the subjects of disagreement are the ways in which
harm should be avoided and who should be responsible
for such avoidance, i.e. questions of justice and
allocation. Before we turn to these questions, we
must however consider the quantitative aspects of
need in relation to health care. Such a topic might
well be viewed as the reality of need provision.

This reality posits scarcity as an absolute
value (see Chapter 1). In other words, scarcity is
taken as given. Indeed, much of the discussion of
need (a 'demand' concept) is given over to an exami-

nation of resource inputs (supply). We recognise, however, that in these examinations there lies a vital consideration - the operationalisation of need. In other words, once a need has been recognised, the courses of action for its satisfaction must be implemented and the criteria for assessing its attainment established. In this implementation and assessment lie the difficulties with 'need'. These difficulties are themselves well-addressed by Butler and Vaile (1984) when they establish analysis at a commonsense level. They state that two separate ideas are expressed in the statement 'this man (sic) needs a doctor.' It implies that his condition is unsatisfactory in some way and should be improved, and that the intervention of a doctor would probably secure the necessary improvement. The statement thus suggests a need for something in order to achieve a change that is regarded as beneficial, a view close to Miller's avoidance of harm.

A similar but more geographical statement is that health region X needs resources. We have of course moved from a comparatively straightforward medical assessment of individual need (although there is medical dispute over a range of needs and treatments e.g. chemotherapy v. surgery for certain carcinomas, abortion v. fostering as a response to 'unwanted' pregnancies) to a social assessment of the health states and provision in regional populations. Such an assessment was the basis of the report of the Resource Allocation Working Party (RAWP) on the regional distribution of resources among English health regions (Great Britain, 1976). The underlying objective of RAWP was to create equal opportunity of access to health care for people at equal risk, to be achieved by selecting "criteria which are broadly responsive to relative need, not supply or demand and to employ these criteria to establish and quantify in a relative way the differentials between geographical locations" (Great Britain, 1976, 7). The success of RAWP and its implications for 'over-funded' regions are immaterial for our present purposes. We should note that need is conjoined with a notion of equality (and therefore justice) and seen in relative terms. Need is relative to what other regions enjoy (and therefore the national average) and to existing levels and types of supply. This second 'relativity' is crucial for need is a quantifiable entity structured in terms of the ways that health care has always been organised and provided in the NHS. Thus as the formula implies, the historic distribution of

supply between the regions is to be overcome but as
the quantitative approach to needs and wants suggests, within the framework of allocation to health
vis a vis other government programmes. Need becomes
a target financial allocation and is satisfied when
actual expenditure is the same as this allocation.
In fact the basic criterion of relative need is
regional population size, weighted for age and sex
structure, SMR, fertility rates and marital status
and adjusted for cross-boundary flows and other
specific costs. The formula applies only to revenue
resources and it may be argued that it is capital
resources that are most crucial as they generate an
ongoing need for revenue expenditure (see Eyles,
1985b). But more fundamental from our present point
of view is whether the formula successfully measures
need. Mortality may not be a useful measure of
morbidity (see Forster, 1977) and SMRs are
themselves subject to random fluctuations if only
small numbers of deaths are involved each year (see
Geary, 1977). Further, the RAWP formula sees need
as a technical judgement, but as we have pointed
out, it is a societal construction, having,
therefore, moral and social dimensions as well. The
formula is also a judgement which sees one outcome
as manifesting need satisfaction, namely the equalisation of revenue resources between regions and
districts for populations at equal risk. But there
may be more than one outcome resulting from intervention.
     Before addressing the issue of relative and
different outcomes, we wish to examine a particularly geographical component of need which arises in
fact from the RAWP formula. It is the notion of
accessibility. The availability and supply of
health care resources do not guarantee their use in
need satisfaction. The resources must be accessible. Accessibility has spatial and social dimensions. Spatial accessibility refers to the physical
proximity of physicians, hospitals, clinics and so
on. In fee-for-service systems of health care, GPs
tend to locate in higher rather than lower status
residential areas (see Lankford, 1971; Shannon &
Dever, 1974; Cleland et al, 1977). Even in
nationalised systems as in the UK, there are
maldistributions (see Knox, 1978; 1979) and in any
event policies to regulate GP distribution and hence
accessibility rarely operate below the scale of
whole towns, so intra-urban disparities remain (see
Knox & Pacione, 1980). Social (or effective) accessibility refers to whether the facility is open at

convenient times, whether it is socially or financially available or whether similar types of consultation are available to all. Social accessibility
depends, therefore, on the nature and quality of
medical practice and the social and economic constraints acting on individuals needing health care.
These dimensions of accessibility (and of utilisation as revealed accessibility) have been reviewed
by Joseph and Phillips (1984). While we should note
that accessibility is a vital element of any allocational system and one to which we shall return when
we consider justice, it requires need to be treated
in rather a specific way. First, supply must be
taken as given if need is to be defined in terms of
relative accessibility. The individual must be
relatively inaccessible to something already there.
It is a nonsense to speak of accessibility in relation to doctors or facilities not there. (It is of
course not a nonsense to speak in terms of needs
themselves in relation to things not present because
needs are goal-related and a goal may be attained by
additions to the social fabric.) Secondly, need
becomes attribute-oriented, a characteristic to be
identified from the social and personal nature of
the individual. The need for health care is seen as
in some way related to the characteristics of the
individuals afflicted by ill-health. Such relationships have been assiduously documented (Townsend and
Davidson, 1982; Eyles and Woods, 1983; Joseph and
Phillips, 1984) but to emphasise them means that
care is equated with treatment and cure (of afflic-
tions) and that need is equated with demand (for
treatment of the affliction).

At least, the emphasis on accessibility (and
utilisation) allow ineffective demand to be treated.
Attributed need is seen as a fair manifestation of
demand for care. The treatment of need and demand
by economists is somewhat differently conceived.
Indeed, Nevitt (1977) has suggested that the term
'need' should be abandoned. Others (e.g. Culyer,
1976) recognise the importance of value judgements
in medical care and that need is itself a normative
concept (see Williams, 1974a). But these recogni-
tions lead to economistic judgements about need and
health care. Williams (1974a) suggests that need is
goal-related but then goes on to see this in purely
technical and economic terms. Needs are, therefore,
manifested in individual preferences (see Lees,
1961) which are seen as deliberate individual
choices (or, of course, demands). If individual
preferences and demands dominate, people are "free

to pick and choose, doctors and hospitals who provided inferior treatment at expensive prices would
lose custom to those who provided better and/or
cheaper services" (Le Grand and Robinson, 1976, 33).
To rebut these arguments, we can do no better than
quote:

> To rely wholly on evidence about short-run
> individual preferences as the criterion for
> collective action is <u>unhistorical</u>, because
> it takes too little account of the way in
> which preferences are shaped (and could be
> reshaped) by influences extending over long
> periods of time. It is <u>unsociological</u>
> because it treats people as atomised indi-
> viduals, deciding for themselves, rather
> than as members of classes, families and
> other groups which support and constrain
> them (and could influence them in different
> ways if the social structure changes
> (Donnison, 1975, 423).

The treatment of need as demand and the insufficient
attention paid to ineffective demand and community
preferences may lead us to treat with caution other
work by economists. To elaborate, we return to the
question of outcomes. There is little disagreement
that need satisfaction may be seen in the improvement in a person's condition. But needs do not
exist in isolation but as part of a matrix of needs,
wants, expectations and desires, which itself is a
cultural determination. But which needs should be
treated and satisfied in a resource-constrained
health care system? In such circumstances, needs
are prioritised and the outcomes of treatment and
intervention assessed for their relative benefits.
As Butler and Vaile (1984, 147) put it, "a judgement
about relative need is a reflection of the value
that is placed upon the expected outcome of intervening in different conditions of need ... Relative
needs are judged in terms of relative benefits."
But not only are there different views as to how and
with what effect a condition may be changed, but
different outcomes are evaluated in different ways.
Economic analysis has again pointed the way in such
evaluations. Indeed, the questions raised in such
analysis are vitally important, but the suggested
economic equation of devoting X% more resources to
improve health by Y% (see Williams, 1974b) seems not
only dubious but dangerous. It is dubious because
it simplifies a complex social and political process, some elements of which, has been addressed by
Rosser and Watts (1974) in their use of court com-

pensation awards to evaluate hospital treatments.
It is dangerous because of its simplifying nature
and its reduction of care to that which can be
quantified. Butler and Vaile (1984) point to the
rough-and-ready hierarchy of outcome evaluations
among those who control the use of resources: in
descending order, conditions which threaten life,
those which threaten functional capacity and those
which cause discomfort. But a cost-benefit approach
to treatment shows that the limits placed on financial resources even to extend life are low. Gould
(1971) calculated that in current prices the decision not to child-proof drug containers implied the
value of a child's life at less than £1000 p.a.
Economic evaluations of outcome in the context of
prioritising as a form of rationing and cost-containment are increasingly meaning upper age limits
for certain forms of treatment. Those making no
current economic contribution to society are given
low priority and treatment may be delayed for months
or years or may not occur at all. It has recently
been suggested that this age limit will be increasingly reduced (and may affect the other dependent
population, children) as the costs of the NHS become
more difficult to meet. We of course do not
'accuse' economists of advocating such a state of
affairs but merely suggest that this is where economic analysis of health care may lead. There are of
course costs involved in meeting needs and expected
costs and anticipated benefits must form part of an
assessment of relative need (see Davies, 1977).
But when need and benefit provide a scant veneer of
justification for a cost-containment policy, then we
must raise questions about the nature and direction
of goals in society. Such questions require a broad
view of need as a normative concept, as a social
construction, and its claim to be satisfied requires
its location in systems of justice.

Justice

Need is in fact only one of the criteria which can
act as entitlement under systems of social justice.
Miller (1976) establishes three - rights, need and
desert - as criteria. His approach attempts to link
social ideas and the social orders that contain
them. He contrasts his approach to that of Rawls
(1972) claiming that "the whole enterprise of constructing a theory of justice on the basis of the
choices hypothetically made by individuals abstrac-
ted from society is mistaken," because people hold

conceptions of social justice as part of more
general views of society and they acquire these
views through their experience of living in actual
societies with definite structures embodying particular kinds of interpersonal relationships (Miller,
1976, 341-2). It may be mistaken but it is not
without merit and Rawls' scheme has certainly
enriched discussions of distributional justice. But
all theories of justice, whether they claim to be
asocial or socially located, use the general criteria and principles of social justice in particular
ways. Some authors in fact attempt to remove the
notion of justice seeing differences between people
not the result of injustice but of natural endowment
(see Acton, 1971; Hayek, 1976). Those suffering
derivation have, therefore, no moral claim on the
resources of society. Welfare is a gift not a
right. A similar notion can be found in Rawls
(1972, 102)

> the natural distribution is neither just
> nor unjust, nor is it unjust that men are
> born into society at some particular posi-
> tion. These are simply natural facts.
> What is just and unjust is the way that
> institutions deal with these facts ... The
> social system is not an unchangeable order
> beyond human control but a pattern of human
> action.

Herein lies the distinction between Hayek and Rawls,
with the former seeing this pattern beyond human
control as there is no agreement over ends, the
latter regarding it as part of our humanness, itself
the basis of agreements and contracts. We accept
Rawls' argument on this point and see justice as a
necessary dimension of human activity. Indeed, we
also agree with Kamenka's (1979, 200) approving
citation of Mill: "justice ... is a name for certain classes of moral rules, which concern the
essentials of human well-being more clearly, and are
therefore of more absolute obligation than any other
rules for the guidance of life." Acceptance of this
comment should not imply acceptance of Mill's indi-
vidualistic utilitarianism. Central is simply
justice as a set of rules or guidelines for human
action. It is not as simple as we imply because
once justice has been elevated to a principle, its
importance as a concept tends to be reduced as the
bureaucratic and administrative calculation of
distribution takes over. As we shall see, all
distributional questions demand administrative
solutions whatever their political pedigree.

Further, as it becomes established a principle of justice tends to shift from 'ought' to 'should', i.e. people ought to be treated equally to people should be equal. This movement itself often fuels further debate about distribution, so that the consensus on the 'ought' breaks down. We may see this as one element in the Conservative realignment of everyday life and social provision in Britain in the 1970s and 1980s.

While we shall return to these themes when we examine systems of justice and allocation, it is sufficient for our present purposes to identify justice as a practical rule-related activity. As Harvey (1973, 15) notes social justice is not merely a philosophical issue but "something contingent upon the social processes operating in society as a whole." It is the principle or set of principles for resolving conflicting claims over the allocation of scarce resources. It, therefore, involves con-crete evaluation, consideration of facts, the selec-tion of principles and descriptions and the ordering of preferences and interests.

Interestingly, Kamenka (1979) examines justice in the narrow sense of these requirements, as law, seeing this institution as performing three related societal functions, namely establishing and main-taining the fundamental rules of living together (peace-keeping and social harmonising function); providing the principles and procedures for resolv-ing disagreements between individuals and groups (conflict resolution function); and guaranteeing and protecting existing productive relations and distributional mechanisms and realising and enforc-ing new procedures (resource allocation function). All these functions are variously included in dif-ferent systems of justice. It is perhaps the last named that is of greatest concern to those inter-ested in geographical analysis, in the distribution and allocation of resources across space and between groups. It is also the most contentious function and in industrial societies the most important in that there is (still) broad agreement on the neces-sity of the other two elements (see also Honoré, 1968). It points up the necessity of rules to govern resource allocations (and indeed social relationships). Law (and justice) are necessary but may be ideological in the specific forms they take. Thus abstract bourgeois law promotes a formal equal-ity which amounts to real inequality in a class society. "The 'Republic of the Market' conceals the 'Despotism of the Factory'," (Pashukanis, quoted in

Kamenka, 1979, 1U). This third function also points
to the tension in modern society between bureaucratic administration and social policy on the one
hand and the law and abstract principles of justice
on the other. It is in the allocation of resources
that the 'battle' over 'who decides', between adju-
dication and administrative fiat, is fought, viz
rate-capping.

The third function is the distributive one and
allows us to extend the discussion to examine distributive justice, the principle(s) on the basis or
bases of which the question who gets what, where and
how may be answered. Influential in examining this
issue has been Rawls (1972) whose work has been
used, among others, by Titmuss (1974), Harvey (1973)
and Smith (1977). Rawls takes people out of their
social roles and has them meet to formulate a social
order (the distribution of relative rewards and
deprivations between social positions) that they
would find acceptable. These principles are the
initial choices of free and rational people, uncon-
strained by what they already possess. As the
social position an individual is to fill is unknown
to him/her, s/he is unlikely to be biased and an un-
biased decision is seen as fair. This somewhat
idealist position (see Plant et al 196U) which turns
on what can be extracted from the principle of
rationality (see Barry, 1973), in the basis of the
theory's development. Rawls (1972, 15U-1) establishes two principles of justice: "a principle
establishing equal liberty for all, including equality of opportunity, as well as an equal distribution
of income and wealth ... (but)... inequalities are
permissible when they maximise, or at least contribute to, the long-term expectations of the least
fortunate group." These ideas along with those of
Runciman (1966) and Davies (1968), have been used by
Harvey (1973) to arrive at principles of territorial
social justice in which need, contribution to common
good and merit are used. It is worth quoting at
some length.

The principles of social justice as they
apply to geographical situations can be
summarised as follows:
1. The spatial organisation and pattern of
regional investment should be such as to
fulfil the needs of the population. This
requires that we first establish socially
just methods of determining and measuring
needs. The difference between needs and
actual allocations provides us with an ini-

tial evaluation of the degree of territorial injustice in an existing system.
2. A spatial organisation and pattern of
territorial resource allocation which
provides extra benefits in the form of need
fulfilment (primarily) and aggregate output
(secondarily) in other territories through
spillover effects, multiplier effects and
the like, is a 'better' form of spatial
organisation and allocation.
3. Deviations in the pattern of territorial
investment may be tolerated if they are
designed to overcome specific environmental
difficulties which would otherwise prevent
the evolution of a system which would heed
or contribute to the common good. (Harvey
1973, 107-8).

It sounds very much like the RAWP formulations now
used in the UK, Australia, New Zealand and the
Netherlands to allocate health care resources.
Indeed, this may be one of the significant problems
with Rawls as the principles are derived in isolation from social context; outcomes are left to fend
for themselves. Rational and consistent judgement
and appraisal are assumed. As Daniels (1975) notes
and the effects of RAWP indicate, there is nothing
in Rawls' institutions of social justice to prevent
the differences in wealth from becoming very exten-
sive. As Harvey (1973) avers, Rawls' formulations
may lead to Marxian or Friedmannite institutions.
Indeed, the institutions which Rawls constructs on
the bases of rational, free actions in a 'veil of
ignorance' are very like those of liberal, welfare-
capitalism (see Hampshire, 1972; Lukes, 1974;
Forder, 1984). This societal location of his analysis may even be found in his term justice as fairness which Kamenka (1979) interestingly notes is an
English construction. In German, Russian and
Chinese, 'unfair' can only be expressed as 'unjust'.
But the implications of his concept of justice are
that the institutions of 'welfare capitalism',
defined in liberal terms, should be reformed to give
justice a higher priority. This point is extremely
important. It may be, as George and Wilding (1976)
suggest, that the values of individualism and equality of opportunity legitimise inequality, but if
justice allows a regard of "the human situation not
only from all social but from all temporal points of
view" i.e. sub specie aeternitas (Rawls, 1972, 587)
then even vested interests might be reluctant to act
in ways that they had been convinced were morally

indefensible. We may note that the Conservative
government's cost-containment policies for the NHS
have still led to the provision of a (somewhat
parsimonious) real increase in resources.

In some ways, justice as fairness becomes a
description of an ideal-type, based largely on the
best features of liberal welfare capitalism. This
comment suggests a relationship, albeit intuitive in
Rawls' context, between justice as an idea and as a
practising social order. We wish to explore this
relationship more fully because it enables the
establishment of criteria to examine systems of
allocation. To investigate systems of justice, we
shall use the ordering frameworks of Miller (1976)
and Lang (1979). Miller's work is important because
it establishes three bases for systems of justice,
namely need, rights and merit. He argues that these
are irreconcilable principles with no overriding
requirement to prioritise them. Each is best understood in relation to different views of society and
is appropriate to different forms of society.

First, rights may be enshrined in law or
defended by moral argument. Their value lies in
their predictability which in turn leads to security
and freedom. Expectations are fulfilled and actions
can be planned within known limits. Rights are
protected by law but also by custom and habit. Such
justice is conservative and exemplfied by feudal
societies with different rights accruing to different positions in the social hierarchy, these being
based on personal allegiance. Secondly, desert or
merit may be based on many qualities. Miller con-
centrates on economic desert and on the effort
expended at work and the contribution made to the
common good by it. Desert is determined by the
market and it is not surprising when Miller illustrates such a system with the capitalist economic
order. In organised capitalism, dominated by the
bureaucracies of business, the state and trade
unions, however, individuals enter most relationships as members of organisations. There is still a
hierarchy of rewards based on talent and socially
defined desert but this is often related to contribution to the common good and to need. (We 'need'
entrepreneurs and property developers more than we
'need' nurses, hence the .justification of inequalities.) Miller is right when he suggests Rawls'
analysis gains much of its power from encapsulating
this synthesis of desert and need. Finally, need is
seen in relation to harm and plan of life (see
above). Justice is based on sympathy, solidarity

and altruism. Miller exemplifies this justice by
communes, including religious communities in fif-
teenth and sixteenth century Germany and the
Kibbutzim in Israel. Communes have in the main been
short-lived and have existed as enclaves in societies organised on different principles. They tend
though to base distribution on need and equality.

It is in this tendency that we may isolate a
problem with Miller's scheme. It also applies to
others. 'Freedom', 'equality', 'fairness', and
'right' do not have unproblematic meanings. Is, for
example, freedom from ill-health equatable with
freedom to choose the health care resources which
enhance well-being? For Hayek (1960), freedom is
equated with individualism where coercion of some by
others is reduced as much as is possible. For
Tawney (1961) it involves the power to control the
conditions of one's own life and as such is insepar-
able from equality, while for Laski (1925, 150) it
can never exists in the presence of privilege "where
the rights of some depend upon the pleasures of
others". Equality may imply simply equality under
the law (Hayek, 1960). There has in fact been
greater dissent among liberals and socialists as
between themselves and conservatives like Hayek and
Friedmann. Titmuss (1974), for example, isolates
four historic principles as they relate to the
individual, namely to each according to his/her
need; to each according to his/her worth; to each
according to his/her merit; and to each according
to his/her work. These themselves raise questions
of definition and we may also note that the last
three are desert principles (see Campbell, 1974;
Plant et al 1980).

Similar questions are raised by Le Grand's
(1982) 'system-approach' to equality, in which he
establishes five types of equality as suitable
objectives for guiding the distribution of public
expenditure. The correctness of such expenditure
and intervention is assumed. He identifies equality
of public expenditure in that provision shall be
allocated equally among all relevant individuals (cf
RAWP programme); equality of final outcome is that
public expenditure favours the poor who have less
private money; equality of use is that those who
need a service obtain necessary treatments (cf NHS
objective: equal treatment for equal need); equality of cost is that costs of consultation or service in terms of costs and time should be the same
for all individuals (the kernel of equality of
access); and equality of outcome, i.e. outcome

associated with a particular service. It should be
noted that neither scheme refers to equality of
opportunity, a term employable by all political
persuasions. Nor is full equality implied. Indeed
even marxist theorists such as Laski and Strachey
reject such a notion in that there is no justice in
equal reward for unequal effort. Further, the
principle 'each contributing according to their
powers and rewarded according to needs' is too
simplistic and unworkable (see George and Wilding,
1976). It assumes individual rather than group or
'average' needs can be the basis of policy. Indeed,
what the problems of definition demonstrate is that
freedom and equality as concepts become conflated
with the practical attempts to achieve them. The
central questions become how much equality, and will
particular policies lead to greater or lesser equality? The ways in which such questions are answered
show the essential reciprocity between ideas and
society. Indeed, this is clearly demonstrated by
Miller's (1976) scheme and the relationship of
rights, desert and needs with Runciman's three
incompatible systems of justice and society, respectively, conservative, liberal and socialist.

This reciprocal relation between justice and
social order is seen in Lang's (1979) critique of
'bourgois concepts of justice'. These conceptions
are seen as constituting the moral justifications of
the mechanisms of the capitalist free market
economy. He suggests that the just distribution is
that which is the result of the proper functioning
of the market, yet there are shifts in the capitalist economy leading to a crisis in economic and
political theory. The corporate phase of capitalism
and the emergence of a capitalist welfare state have
led to different conceptions of justice. Yet all
are ideological in terms of their relations to the
basic structure of society. All are conformist in
that they "directly or indirectly accept the capitalist economic structure based on private ownership
of the means of production" (Lang, 1979, 127). The
conceptions do, however, differ with regard to their
evaluation of present state of inequalities, their
views on the need for changes in the distribution of
national resources and on the use of extra-market
mechanisms (social policy) to effect redistribution;
and their judgements on .the efficacy of such mechanisms. Lang suggests a threefold classification of
conformist theories, namely reactionary (those
desiring greater inequality), liberal conservative
and progressive reformist.

Liberal conservative conceptions can be seen in the works of Hayek (1944, 1960) and Nozick (1974). For Hayek, justice is an ethical standard which can only be applied to the behaviour of individuals. Any planning or social engineering are threats to a free and open society. In a capitalist system, social justice is a dangerous mirage. The market is a spontaneous force and the distribution of goods therefore produces no ethnical questions (see Hayek, 1966). The relevant type of justice (commutative) enschrines the principle that services rendered should be rewarded according to the value they have for the recipient, irrespective of the needs or deserts or characteristics of those rending the service. Commutative justice ensures freedom of choice of those who are parties to transactions. A similar view is developed by Nozick (1974) who regards justice as entitlement, so that a just state is obtained if people own or hold the things they acquire legitimately. Everyone is expected to do what s/he decided to do and everyone ought to receive what s/he has done for him/herself and what other people have done for him/herself. In such conceptions, liberty is given priority over equality and economic efficiency over justice. The only relevant differences between people are those required or produced by the market or determined by the legal procedures for acquiring goods. Charity is seen as the most proper way of helping the poor and ill while less extreme versions suggest a safety net of services might be provided for the inadequate. Economic growth is seen as eventually providing enough goods to distribute. That distribution will still be a matter for the market.

Progressive-reformist views of justice stem explicitly from the previously discussed work of Rawls (1972). Such views have as a central premise the need for state intervention to correct inequalities inherent in the capitalist mode of economic organisation. George and Wilding (1976) separate these interventionists into three groupings: the reluctant collectivists, the Fabian socialists, and the marxists. The first - Beveridge, Keynes, Galbraith - aim to purge capitalism of its inefficiencies and injustices so that it may better survive. Beveridge (1943), for example, wanted to use the power of the state to abolish want, disease, ignorance, squalor and idleness which were seen as common rather than individual enemies, while Galbraith (1967) saw the need for the state to manage aggregate demand and ensure minimum incomes

to alleviate poverty and illness. Indeed
Galbraith's view of the relation of capitalist
economy and state is a benign mirror view of
Habermas' (1976) interpretation of the state ensuring the conditions of and material infrastructure
for capitalist accumulation. For the reluctant
collectivists, though, poverty rather than injustice
was the major problem.

Injustice was recognised as a powerful inequity
by the Fabians. Intervention was necessary to
modify injustices of the market system of distribution (see Tawney, 1964; Crosland, 1956). Welfare,
justice and equality were issues of collective
responsibility. Society as a whole should assume
the costs of social and economic development so as
to share them as equitably as possible among all
members. Social policy is not however just a matter
of ameliorating economic misfortunes it is also an
agency of social integration,, binding an aggregate
into a community. Titmuss (1968) was surely correct
when he said social policy has concentrated too much
on the poor, on being an ad hoc appendage to economic growth and the provision of benefits rather than
the formulation of rights. The differences between
the Fabian view and the reluctant collectivists is a
matter of degree, manifested in practice, rather
than one of substance. Equality (differently conceived - of use or access for the first group; of
outcome or 'opportunity' for the Fabians) as well as
liberty is seen as important and justice as fairness
(and based on the rights of citizenship) must be
considered as well as efficiency (their relative
weighting depending on national economic performance). Both are conformist in Lang's sense,
regarding the capitalist system as the best means of
obtaining the necessities of well-being and/or
amenable to change to reduce poverty, ill-health and
injustice.

With George and Wilding's third category - the
marxists - we engage Lang's (1979) non-conformist
conception of justice - the marxian. The theorists
discussed by George and Wilding see the abolition of
the capital system as a precondition for the removal
of social ills. Such discussions have established a
powerful critique of bourgeois theory and practice,
but little construction has occurred. This is in
part corrected by Lang (1979; see also Hancock,
1971). The two principles - from each according to
their capacities, to each according to their needs -
are examined. It is pointed out that there cannot
be unlimited need satisfaction, individual opportu-

nities being limited by genetic endowment and societal ones by technological, cultural and economic
developments. Desert must form part of social
justice but in accordance to contribution to society
through work. Further, justice is seen as part of a
morality which not only defines the claims of individuals in relation to society but also the morally
justifiable rights of society concerning individuals, i.e. their duties. We should note that the
rights of the feudal lords carried with them the
responsibility for the welfare of their subjects.
Rights and duties are of course subject to manipulation and redefinition in practice. That has indeed
occurred in state socialist societies, where the
rights of individuals as individuals are considered
insignificant. Indeed, the strength of such societies is their provision of certain collective
entitlements usually with respect to objective
living conditions. Basic needs are satisfied
through egalitarian state provision. This is in
fact Lang's first element of providing a minimum of
services for a just society. Indeed, his conclusion
of his discussion on a marxian conception of justice
is disappointing, advocating "minimal equal standard
of living for all citizens regardless of their
individual desert," especially with regard to
natural environment, housing, health services and
primary education; limiting "the gap between individual incomes" by law, although desert remains
important; and ensuring "for disabled people suitable living conditions corresponding with the
general standard of living of the society" (Lang,
1979, 123). With the addition of limiting income
differences, it sounds remarkably Rawlsian. It also
has parallels with Stretton's (1976) interesting
scheme for greater equality in a world of ecological
scarcity. The convergence may be due to the fact
that all systems of justice have to be put into
practice and there may be limits to the number of
ways of providing goods and services and avoiding
harm and bads. Even, the marxian conception, to be
of any practical meaning, has to be operationalised.
The collective assessment of the worth and desert of
work and individuals has to be carried out. Services have to be collectively provided. Such a
conception demands therefore central planning for
equivalence of assessment and then the establishment
of norms for services provision and rates for
employment. With central direction, it is however
perhaps easier to decide the type of accessibility
established in the system in terms of centralised or

diffuse services. But as with all the systems of justice, our discussion tends towards a considera- tion of the practical, of allocation. As we have been at pains to point out, there is a reciprocal relation between a system of justice and a social order. That relation is manifested in the ways in which a society allocates resources to its members and across its territory.

Allocation

In our discussion of systems of justice, three broad types were identified: the liberal-conservative, progressive-reformist and marxian-socialist. We have already noted how they relate in respective principles of justice rights, desert and needs. We must point out that the equivalence between cate- gories is not perfect. A liberal-conservative system is as concerned with deserts as with rights, and desert enters a marxian system. Further, the three types of allocative mechanism we respectively identify as the market, bureaucratic intervention and central planning are not the province of one particular system of justice. The categories should be viewed as ideal-types, more or less linked to one another and across categories. These linkages are heuristic devices which break down when we begin to consider the specificities of particular cases. The broad relationships are established in Figure 2.1 which can in fact be seen as a summary of this chapter.

The market mechanism is closely allied to the liberal-conservative view of society. The market is seen as "the superior means of registering prefer- ences" (Lees, 1961, 14). Other means of allocation are not only wasteful but also wrong. They are wasteful because they provide welfare services for zero price at the point of use and may stimulate false and insatiable demand. Waste is also found in inefficiency. State monopoly of provision leads to unnecessary expenditure because of insufficient attention to costs and efficiency. Competition, on the other hand, means not only better and more economic use of resources but also to experiment to find better ways and methods of distribution. They are wrong because of their views on human nature and their repercussions for human relationships. Health care and other forms of welfare promote social disruption by recognising that individuals have rights based on their citizen- ship. The "translation of a want or need into a

Figure 2.1 Need, justice and health care

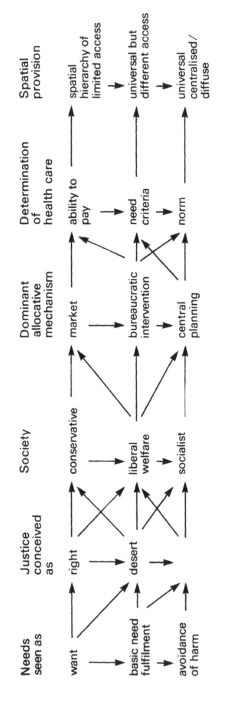

Note : As the text makes clear, the categories are not absolutes but tend to merge one with another. Further, the relationships shown are only possibilities : they are not one-way; neither do they suggest any necessary progress or development. For example, just as liberal–welfare societies tend toward socialist principles at some times, at others they tend toward conservative ones.

right is one of the most widespread and dangerous of modern heresies," providing "unlimited fuel for dissatisfaction and unlimited scope for the fostering of animosities between one section of potential recipients and another" (Powell, quoted in George and Wilding, 1976, 27). These rights and grievances in a framework of little agreement on means or ends entail the imposition of social goals and distribution by the state. This inexorably leads to totalitarianism, dictatorship by the state. "Planning leads to dictatorship because dictatorship is the most effective instrument of coercion and the enforcement by ideals" (Hayek, 1944, 53).

Rather, what is necessary is the advocacy of "the use of price and market mechanisms and competitive forces, based on an underlying philosophy which starts from a strong preference for the decentralisation of initiative, and for the revival or extension of freedom of choice for individuals as buyers and sellers or consumers and producers" (Hutchinson, 1970, 10). In this, the key value is that of freedom. Freedom of the individual or family is seen as the ultimate goal in judging social arrangements. This freedom allows the individual to make the most of his/her capabilities and opportunities according to his/her own rights (Friedman, 1962). Freedom is seen as freedom from undue interference by others. The voluntary association of individuals is achieved by "the technique of the market place" (Friedman, 1962, 13). Further because we can only know about a small part of society, the market is the only way of ensuring that every individual uses his/her unique knowledge of particular opportunities to achieve an overall social order (Hayek, 1968). The market is the only way of regulating the world, especially as with society itself being unknown, it is logically impossible to conceive of a socially just order (see Hayek, 1976).

The market grants freedom in the disposition of individual skills and property. That freedom to choose how to dispose of skills brings with it a responsibility for his/her own behaviour and its consequences. It is their responsibility to provide for their own welfare. We may wish to help others but that is our individual choice. It is "an obligation towards people of one's own choice, it cannot be enforced under equal rules for all" (Hayek, 1976, 765). Charity, or voluntary contract (see Nozick, 1974), becomes a basis of welfare. Those advocating the market principle of allocation are thus against the state provision of 'free' health services. It

encourages a dependence that might lead to totalitarianism and 'serfdom'. People must be encouraged to
help themselves and the state's main role is seen as
encouraging or enforcing individuals to insure
themselves against sickness with private insurance
companies. Such encouragement may be seen as part
of what Titmuss (1974) called the 'industrial
achievement-performance model' of social policy,
whereby needs are met on the basis of merit, work
performance and productivity. It is assumed that
individual incomes reflect sufficiently their merit,
performance and productivity and that is the role of
the state "to support industrialism and the attempt
to establish a completely competitive, self-regulating market economy founded on the motive of individual gain" (Titmuss, quoted in Room, 1979, 52).

Some market liberals, however, see the role of
intervention as being somewhat larger than that just
implied. Friedman (1962) in particular sees the
necessity for intervention in, for example, rule-
making and in cases where voluntary exchange is
costly or virtually impossible. More problematically, he accepts intervention on the grounds of
paternalism for those designated as not responsible.
Clear examples, given his unproblematic definitions,
are the severely mentally ill and the severely
mentally handicapped. For others - children, minimum financial aid to the destitute - the criterion
of responsibility is not the sole one. He is forced
to accept that responsibility is taken from the
individual and others have to decide by "a consensus
reached by imperfect and biased men through free
discussion and trial and error" (Friedman, 1962,
34). Intervention though, on the grounds of
charity, paternalism and compassion, provides a
minimum, a safety-net below which no citizen should
be allowed to fall. Minimum income, food stamps,
educational and health care vouchers may be used.
Their provision is vigorously means- or income-
tested and aimed at individuals. "The subsidisation
of institutions rather than of people has led to an
indiscriminate subsidisation of all activities
appropriate for such institutions, rather than of
the activities appropriate for the state to sub-
sidise" (Friedman, 1962, 100).

Welfare becomes residual. It is means-tested
and for value-for-money locally administered because
those in localities better know the needs and cases
of individuals in distress rather than central
authorities. The level of care obtained by an
individual depends, however, largely on his/her

position in the labour market and the level of
income achieved. In short, it crucially depends on
ability to earn and ability to pay for services.
Access to care is limited by financial (income)
constraints yet the hierarchy of facilities in such
a system will be a complete one from primary to the
most sophisticated types of secondary and post-
operative care. This hierarchy will be privately
provided and its distribution in space will be
oriented towards the wealthier parts of a country.
Poorer parts receive rudimentary care in terms of
fewer general practitioners and low-funded public
hospitals which may be supplemented by voluntary
provision from charitable organisations.

The American health care system is often taken
as the illustration of allocation by market principles and of the differential spatial provision and
access. We wish, however, to refer to the British
context before the inception of the NHS. This
context ably demonstrates the unequal outcomes of
market provision and the interplay of charity,
minimum public provision and the encouragement of
self-help. The reform of medical education compli-
cates the picture at the end of the nineteenth
century, but by 1866 there was 1 doctor to 1,662
people in England and Wales. But as F.B. Smith
(1979, 368) comments:

the distribution of GPs generally increased
their scarcity value. In 1861, the census,
relying on a wide definition of 'doctor',
reported 1 per 514 persons in London,
compared with 1 per 1,769 in Wales. By
1866 only two 'unhappy' places were
reported to be over-supplied: Brighton
with 1 per 726 and London, with 1 per 939.
Below these, the distribution ranged from
Bristol 1 per 1,232, Liverpool 1 per 1,564,
Glasgow 1 per 2,269, Sheffield 1 per 2,593
to insalubious Salford with 1 per 3,968.
But these averages must hide grossly uneven
distributions between rich and poor places
within cities. In 1914 Shoreditch had 1
per 5,666 compared with Kensington's 1 per
566.

These figures do give some idea though of the impact
of a system in which GPs acted like any other
retailer, setting fees the market could bear and
being in direct competition to one another. In some
areas, this private system was never workable so
workers, as in the Northern coalfields, banded
together to hire a doctor, this club system being an

important antecedent of the NHS (see Widgery, 1979).
Access to specialist opinion was limited and so
therefore was the quality of care. Doctors were
also employed under the provisions of the Poor Law
but, on the grounds of individual responsibility and
less eligibility, only attended the utterly destitute. For those seriously ill, a transfer to the
workhouse or Poor Law infirmary was possible.

The Poor Law infirmary provided the bulk of
secondary care. In many areas it was the only
source of care. In larger cities there was a choice
between these institutions and the voluntary hospitals, i.e. charitable institutions. Both came to
introduce charges for treatment and screen patients.
The Poor Law infirmaries submitted their patients to
a means test and had separate services for those
unable to afford fees (see Abel-Smith, 1964;
Walters, 1980), while the voluntary hospitals adopted a system of payments, such that in London such
income accounted for 10 per cent of the total in
1920 and 50 per cent in 1939. In fact, "by 1939,
the majority of hospital patients paid for their
treatment, either through insurance schemes or by
arrangement with the hospital almoner. Only the
very poor could be assured of free treatment, and
even this was dependent upon the discretion of the
hospital involved" (Stevenson, 1984, 214).

This social accessibility problem was compounded
by the spatial maldistribution of facilities,
brought about by the operation of the market allocational mechanism. By 1882, in London the great
population movement to the south of Guy's Hospital
had no nearer hospital to treat them while only a
small hospital at Dalston served the new northern
suburbs. "Ten of the 15 large metropolitan hospitals with three-quarters of the beds, were within
one and a half miles of Charing Cross. Two, the
London with 790 beds and the Metropolitan Free with
20, had to cope with the East End with a population
of over one million. One, the Great Northern with
33 beds, had to serve north London with a population
of over 900,000" (F.B. Smith, 1979, 251). This
maldistribution might be simply put down to the fact
that there is infrastructural inertia. Hospitals
cannot be relocated as fast as people. This is of
course true but the movement of institutions that
was suggested was for financial or doctors'
convenience reasons not for patient care. The
Westminster Hospital Board discussed moving from
Petty France to Charing Cross because the latter was
a wealthier district and the former had only poor

subscribers. The proposals to move the London
Hospital to a less congested and more salubious site
were blocked by medical staff not on the grounds of
patient accessibility but their own commuting
arrangements.

These patterns persisted into the inter-war
period. The London region had 10.2 beds per 1000
population compared with 4.9 in South Wales. Great
disparities also existed within regions with
Yorkshire's 6.6 per 1000, masking 12.3 in Dewsbury,
10.8 in Halifax, 4.9 in York and 3.1 the North
Riding (Walters, 1980). There was a lack of consultants and qualified specialists outside London (see
Titmuss, 1976) and some of the worst medical care
was found in rural areas. Indeed, hospital care
generally remained largely uncoordinated despite the
increasing intervention of the state. These interventions were mainly enabling rather than directing.
Most local authorities preferred to make grants to
voluntary hospitals rather than establish public
ones. Iliffe (1983, 21) comments that "the hospital
system was ... chaotic ... The quality and quantity
of hospitals in any area, and the services they
provided, depended on the interests shown in them by
local government, by charitable organisations and by
enthusiastic reformers working within them". Great
changes did occur in the inter-war period especially
in maternity and child care but administrative chaos
and economic crisis meant that there was little
concrete intervention (see also Bruce, 1971). In
many ways, this period was a battle ground between
groups supporting different allocational principles
- the market and bureaucratic intervention, with
circumstances often conspiring against the latter.

Indeed, this period saw what many regard as the
incipient stage of the welfare state with the passing of the National Insurance Act of 1911. We may
regard this as a movement from encouraging to enforcing individual responsibility and self-help. While
as Doyal (1979) points out this Act was designed to
assuage working class demands and improve 'national
efficiency', it provided a basic income during
illness and a national scheme of primary medical
care. By 1913, it covered about 12 million workers,
increasing to almost 20 million by 1938 and over 25
million by the mid-1940s. The scheme provided a
free GP service and sickness benefit paid through
'approved societies'. It was not universal and did
not cover the dependents of insured workers, most of
the unemployed, most of those in non-manual occupations and all those requiring hospital treatment.
For those it did cover, it established the panel

system: the nationalisation of the club system.
But there were many complaints that 'panel patients'
were often less adequately treated than fee-payers.
Indeed private insurance and sick clubs were the
major sources of care provision for those excluded
from the provisions of the Act, although this proportion declined from about two-thirds to roughly
one-half. Those too poor to participate in the
contributing scheme were forced to pay for a doctor
as they used him/her or attend the outpatients'
department of a 'free' hospital.

The scheme was very much a half-way house in its
organisation and outcome. The 'approved societies'
included not only trade unions and friendly societies but also commercial insurance companies. Doyal
(1979) has pointed to the role and significance of
those companies, showing how the visit to a home for
the giving out of public benefits enabled the sel-
ling of private insurance too. This was necessary
for 'additional benefits' e.g. dental and ophthalmic
treatment. The number of approved societies grew
greatly and by 1939 over 7000 were involved, making
administration of the scheme fragmented and competitive. Further in its organisation, doctors were
given freedom of choice whether they joined or not.
Also important was the maximum income level for
individual participation in the scheme. This was
fixed low enough to exclude all patients just able
to pay for treatment. Doctors were free to choose
how they practised and able to increase their
incomes significantly. This freedom and ability
were heightened by the spatial preferences of doctors, these also being in part market-derived. They
preferred the higher fees and more congential working conditions of private practice. They wished to
avoid the isolation of rural practice and the confines of the club system in working class areas.
Thus middle class areas were likely to have a sur-
feit of GPs while working class areas were under-
doctored. In the late 1930s, there were 50 per cent
less doctors per capita in South Wales as in London
and only 25 per cent as many per capita in the
industrial Midlands as in Bournemouth. Hastings has
one doctor per 1178 population, Swindon 1 per 3100,
Greenock 1 per 3500 and South Shields 1 per 4100
(Leff, 1950; Walters, 1980). But the end of the
pre-NHS period more than most demonstrates how the
abstract allocational principles become re-shaped
and re-defined in practice. The market and bureaucratic intervention were intermingled as the pressures for reforms of the organisation of medical

care became louder from the Labour movement wanting
a universal and unified system of free provision;
doctors with urban, working-class practices with low
salaries; and the middle-class who were worried
about the income limit for national insurance
schemes and could not qualify as charity patients in
voluntary hospitals. Bureaucratic intervention
became more ascendant.

In the second principle of allocation, bureaucratic intervention, we treat together those considered by George and Wilding (1976) to be reluctant
collectivists and Fabian socialists and by Room
(1979) as liberal pluralists and social democrats.
Such distinctions, while conceptually important,
become in practice differences of degree rather than
substance. The differences then lie in the reasons
for intervention not in the commitment to, or belief
in the efficacy of, intervention. They are all concerned in Mishra's (1981) phrase, with 'mending the
world'. This mending is necessary because economic
development will not abolish poverty and ill-health
as natural outcomes of increasing prosperity.
Capitalism is not seen as a self-regulating force.
Inequalities arose from it and often deepened. The
reluctant collectivists, however, continue to see
capitalism as the best system but one requiring
regulation and control if it is to function effi-
ciently and fairly. The state is thus seen as an
independent phenomenon standing in judgement on the
functioning of the economic system, the performance
of private enterprise and the distribution of
national income and resources. It grants citizenship and through citizenship specific rights to
enjoy particular levels of material well-being and
health and welfare care. As such, the state not
only regulates the capitalist system but integrates
the population, promoting an identity of interest
among all social groups (see Room, 1979). This view
of welfare is close to that advocated by Marshall
(1965) and described in Eyles and Woods (1983,
chapter 2). It was the view of health and welfare
that underpinned the Beveridge reforms. It legiti-
mises the institutional model of welfare, seeing the
provision of universal, comprehensive services as a
right. It also is important in the de-stigmatisa-
tion of provision, which is an entitlement rather
than societal largesse.

Intervention is necessary but should be limited.
The market is seen as the source of initiative and
bastion of freedom. Intervention must be problem-
oriented rather than promotional. Its role is to

react, to abolish avoidable ills. The intervention
should be comprehensive but minimal with the provision of benefits above this level being an individual responsibility. Intervention is a matter of
necessity, rather than principles. Galbraith (1958;
1974) argued that the affluent society increased the
need for intervention to care for the children of
working mothers, to combat lawlessness, to provide
medical treatment for the health problems generated
by affluence (e.g. obesity, cirrhosis, lung cancer,
heart disease, nervous disorders). Intervention
will remove the inefficiencies and injustices of
capitalism so that it might survive.

The Fabian socialists support intervention as a
point of principle, regarding it as the only means
by which equality, freedom and fellowship may be
instigated. With this instigation, harmony,
efficiency, justice and individual self-realisation
may be realised. This cannot occur in market
systems as they are geared to effective demand and
not need. Such a gearing will increase and not
reduce the inequalities on which conflict, injustice
and alienation breed. Great emphasis is therefore
placed on state intervention through its political
and administrative mechanisms. Tawney (1961, 16C)
argued that "society is not an economic mechanism
but a community of wills which are often discordant,
but which are capable of being inspired by devotion
to common ends". The state is the way to remake
society and indeed the changes instigated in the
immediate post-1945 period led those like Crosland
(1956) to regard Britain as no longer a capitalist
society and to see the mixed economy as the way to
achieve greater equality. Intervention must be
purposeful and direct with the state having direct
responsibility for welfare, for modifying injustice
in the market system of distribution, for informing
all where the public interest lies, and crucially
for ensuring the economic growth necessary for the
redistribution of resources. The last mentioned is
crucial as the approach is gradualist and reformist
and dependent on new financial resources to meet new
claims. The concern is of course with welfare
rather than equality. The ends of intervention were
the enhanced incomes and health of the disadvantaged
rather than reduction in inequalities. It would
also ensure social integration by seeing provision
as a right (see above). But as Titmuss (1965) has
pointed out, it has been assumed that intervention
and legislation led to provision and that this
provision was equally accessible to all. The lack

of clarity on objectives meant there has been little attempt to co-ordinate policies - one of the major themes of this book - particularly as Crosland (1974) points out for deprived inner city areas. Principled intervention regards the particular administrative method as important and tends to neglect the ends of welfare (Rein, 1970).

Of course administrative method is important. In complex societies, all attempts by public and private bodies to enhance health and well-being require an administrative apparatus. We thus regard the criticism of bureaucratic intervention in terms of it being an administrative solution to social problems as wrong-headed. It must be administrative, although the solution need not take a particular form. It may however be suggested that by its nature bureaucratic intervention tends to be centralised and to involve instruction and policy directives from the top to the bottom and from core to periphery. For this reason, the practices of interventionists of different types become similar. This is particularly true over time when the bureaucracies administering policy develop their own values, outlooks and interests (see Eyles, 1985b). Policy-making and intervention become administration-rather than need-oriented, concerned with change at the margins rather than radical overhauls or shifts in direction. They become elements of 'bureau-incrementalism' (Walker, 1984, 71). Policy is adjusted incrementally in the light of experience and subject to continued agreement. It is, as Lindblom (1959, 66) puts it, "a process of successive approximation to some desired objectives in which what is desired itself continues to change under consideration".

Such bureaucratic intervention may be illustrated by reference to the development of the welfare state and the NHS in Britain. It is not our intention to rehearse the history of either entity (see Gough, 1979; Klein, 1983). Its aim was established in the 1944 White Paper: "everybody in the country should have equal opportunity to benefit from medical and allied services" (quoted in Allsop, 1984, 39). The impact of the NHS was dramatic.

> Within a year of its inception 41,200,000 people were covered by the National Health Service. 'Workman's Insurance' had gone national, paupers were now citizens, what had been grudgingly given as assistance was there as a right. The sign of people trying on spectacles in their local

> Woolworths, the clatter of ill-fitting
> dentures and the hateful sight of nurses
> and sisters selling flags and collecting
> money were abolished. In the first year
> 187,000,000 prescriptions were written out
> by over 18000 general practitioners,
> 8,500,000 dental patients treated and
> 5,250,000 pairs of glasses prescribed"
> (Widgery, 1979, 28).

Access became a universal right and similar types of
care were available all over the country, although
certain specialisms remained concentrated in major
teaching hospitals. This universal social and
spatial accessibility was aided up to the mid-1970s
by growing social expenditures. Navarro (1978)
shows that public expenditure rose from 13.5 per
cent of GNP in 1913 to 40.2 per cent in 1948 and
52.1 per cent in 1968 while social expenditure
(including all public housing monies) rose from
about one-third to one-half of public spending over
the same period. He notes that this growth is
related to the great increase in public employment
with two-thirds of all public expenditure being the
wages of state employers. Gough (1979) demonstrates
that the period 1965-75 showed a great increase in
real terms on public spending on services with NHS
expenditure increasing by 70 per cent. This growth
in expenditure has slowed dramatically since the
mid-1970s with the absolute increases between 1979-
80 and 1985-6 being of the order of 16 per cent.

Not all regard the intervention of the NHS as
beneficial. Walters (1980) sees little change in
the use of service rates between 1947 and 1951.
This may however be too short a period for an adequate assessment. It also misses the point about
the greater availability of services and the fact
that they were available on a different basis, i.e.
they were free at the point of delivery. The NHS
did increase the accessibility of health care especially for women. GPs were also more equally distributed throughout the county. In fact, the early
years of the NHS despite being difficult financially
because of the economic climate of the country did
improve medical services considerably. In 1958, the
Minister of Health said that "Effective beds are up
6 and a half per cent; in-patients admitted are up
by 29 and a half per cent; the ratio of treatment
to beds is up by 22 per cent; new outpatients
treated are up by 12 per cent, and the waiting lists
are down by 11 and a half per cent" (quoted in
Klein, 1983, 31). But such improvements were bought

at relatively higher costs and the problem of
'indefinite demand and finite resources' was to
bedevil the NHS despite the fact that the Guillabaud
Report in 1956 showed that the devaluation of sterl-
ing and the miscalculation of basing initial demand
on pre-NHS utilisation patterns were the causes of
high costs rather than unchecked spending (see
Iliffe, 1983).

But financial strategy meant that the Ministry
of Health decided upon control of inherited local
health authority budgets rather than devising an
appropriate financial allocation for distributional
needs. An administrative solution was therefore
couched in terms of bureau-incrementalism. This
solution tended to perpetuate existing provisions.
In 1950 Sheffield RHB had 9.4 beds per 1000 population while South-West Metroplitan RHB had 15.1. In
1960, the equivalent figures were 9.1 and 14.2
respectively (Klein, 1983). Further, the improvements in the distribution of GPs began to break
down. The NHS had introduced a system of negative
controls with GPs prevented from practising in the
relatively well-doctored parts of the country. The
proportion of patients in under-doctored areas
(those where list sizes were exceptionally large)
fell from 51.5 per cent in 1952 to 18.6 per cent in
1958. This fell further to 17 per cent in 1961 with
the greatest improvements being found in South
Wales, Northern England and the Midlands. The
percentage in underdoctored areas increased, however, to 34 per cent in 1967 (Hart, 1971). This
reversal was caused mainly by the declining number
of medical graduates and increasing deployment to
hospital services. Indeed, Navarro (1978) shows how
great the variations in average list sizes were in
1973, ranging from 1100 in the Isles of Scilly in
the South West to 2951 in Hartlepool in the North.
He suggests that this reversal has been caused by
the reestablishment of the dominance of the professional elite- the London-based consultants - which
strengthens the trends towards urbanisation and
hospitalisation inherent in the capitalist system.

The dominance of the hospital sector cannot be
denied. Nor can the significance of professional
power. But also implicated must be the changing
nature of bureaucratic intevention. The emphasis on
the control of finance saw the 1960s and early 1970s
as being marked as the period of efficiency and
rationality in the use of resources and the reap-
pearance of market principles in the resurgence of
private sector medicine. Efficiency and rationality

are important dimensions of bureaucratic behaviour
and must be seen against the backdrop of national
economic and NHS administrative problems. Consensus
over means - the commitment to the NHS - ruled out
many policy options, despite there being no clearly
stated objectives about what the service should
achieve except in the most general terms. But
attempts were made in the 1962 Hospital Plan to
formulate national standards based on the district
general hospital which was initially to serve 100000
to 150000 and later 200000 to 300000 people. Mohan
(1984) has shown, for North East England, how the
Plan implied a considerable spatial concentration of
services in the interests of the efficiency of the
system as a whole. Economic problems led to the
deferment of schemes and its concentrating tendencies coupled with the dismantling of rural public
transport services deprived people of the means of
access to services.

But overall the Plan became negotiable in the
spirit of 'muddling through' or flexibility in the
light of changed circumstances and local needs.
This subversion of the Plan meant that the hoped-for
redistribution of resources failed to occur. The
building programme was too susceptible to the
vagaries of the economic climate to help to achieve
equity in geographical distribution. This type of
planning was, therefore, superseded by a strategy
for sharing out fairly available revenue resources.
As Klein (1983) comments this is essentially a
rationing strategy in that unlike the planning
approach it did not make any assumptions about the
desirable level of provision but only about the
equitable share of resources between regions and
sectors (cf RAWP). Further, problems of provision
became seen as problems of the service itself,
amenable to rational solution by changes to administrative structures. In many ways, this may be the
final stage of bureaucratic intervention and the
ultimate state of bureau-incrementalism. Policy
becomes not just administration-oriented but admini-
stration-structured. Changes in management structure are seen as the way of effecting a more
rational, efficient and equitable service as with
the NHS reorganisations of the 1974 and 1982.

Part of the rationale of the reorganisations was
to decentralise services, the importance of which
has been shown by the fight to keep open small
hospitals (Ham, 1981; Mohan, 1984). Thus while the
early phases of NHS intervention attempted to create
universal provision and accessibility with decen-

tralised GP services and centralised hospitals the
more recent phase has emphasised a more diffuse
system for all types of care. Cost-containment has,
however, meant that this care is more likely to
stress 'care in the community', prevention and
groups practices-cum-minor surgeries than the
re-establishment of 'cottage' hospitals. The
effects of this cost-containment are compounding the
problems of NHS provision. The differences in
regional resource allocations remain (see Chapter
1). There remain class differences in the actual
and effective use of all services (Walters, 1980;
Doyal, 1979). These continuing problems are
eloquently summarised by Hart (1971, 412).

> In areas with most sickness and death,
> general practitioners have more work,
> larger lists, less hospital support and
> inherit more clinically ineffective tradi-
> tions of consultation than in the healthier
> areas; and hospital doctors shoulder
> heavier case loads with less staff and
> equipment, more obsolete buildings and
> suffer recurrent crises in the availability
> of beds and replacement of staff. These
> trends can be summed up as the inverse care
> law: that the availability of good medical
> care tends to vary inversely with the need
> of the population served.

A major cause of this state of affairs may be
the allocational principle of bureaucratic intervention itself. It is gradualist, seeking to modify
the effects of market forces by replacing them in
one arena of life - medical care. It cannot and
does not do much about the effects of such forces on
income distribution and class life chances. It has
not been clear about its objectives, assuming that
the post-1945 health and welfare changes were a
social revolution rather than an administrative
tidying up of a material benefits system (see
Titmuss, 1965). It does not seem to know how needs
and resources may be matched. And it has become a
phenomenon with its own values, aims and outlooks.
The increasing bureaucratisation of NHS intervention
has loosened its links with the wider social world.
But this criticism should imply that bureaucratic
intervention is not a sufficient condition for
realising 'equality'. It may well, however, be a
necessary one: not only in terms of all solutions
being of necessity administrative, but also in term
of the practical achievements of the NHS. We concur
with George and Wilding (1976, 84) themselves cri-

tics of the welfare state. "The welfare state plays
a key role ... educating the public and the politicians about the continued existence of avoidable
individual and social ills, all the time creating
and inculcating new standards of welfare ... It is
... only the limited and partial achievement of some
socialist goals ... a power ally raising aspirations, widening reference groups, illustrating and
exacerbating the value conflicts of welfare capitalism and providing a dynamic for further change
... a stepping stone toward socialism."

While need criteria influence allocation determined by bureaucratic intervention, they are also
present in a socialist system of allocation, forming
the bases of the norms which ensure equivalent
distribution. But needs and norms are treated in
particular ways. We have already noted how need is
seen in relation to the avoidance of harm. It is
also viewed with respect to social and human potential, especially by marxian socialists. Heller
(1976) notes that Marx distinguished between conscious needs (those relating to personal consumption) and unrecognised needs (these becoming represented by the communal satisfaction of needs). In
a future society, these latter needs which are only
satisfiable socially appear as conscious and personal needs. Their satisfaction will be seen as of
such significance that people themselves set limits
to other needs. Thus is a society of 'associated
producers' only other needs can set limits on human
needs. Thus all needs are those of individual
people and it is their social arrangements and
societal goals that determine which needs are
'dominant' and how they will be satisfied. In a
socialist society in which the realisation of human
potential and the self-realisation of human personality are overarching, communal need and communal
satisfaction will dominate. How such a society is
realised practically as well as theoretically
remains problematic.

The evidence of such state socialist societies
as the USSR and Eastern Europe is not encouraging.
The harmony of interests and integration of institutions are not present (see Lane, 1976; Mishra,
1984). But such societies do produce norms for the
minimum standards of care and resource provision and
have make great improvements to the health status of
the mass of the population compared with pre-
'socialist' times (see Eyles and Woods, 1983, Chapter Six). The critique of the welfare state under
capitalism by marxists does, however, fail to recog-

nise that the utopias that they wish to create do
not exist in any society. The marxist-socialist
critique has, at the same time, to attack and defend
the welfare state and health care system. It has to
demonstrate the role of the state in guaranteeing
the preconditions for the accumulation of capital
and conceal these purposes in a legitimising function of which welfare is a part. O'Connor (1973)
sees accumulation and legitimisation as the two
major functions of the state in capitalist society.
These were seen as conflicting aims likely to result
in a fiscal crisis. But the aims may be mutually
supportive with accumulation (as promises of returns
from economic growth) being an important source of
legitimisation for the capitalist system. Gough
(1979, 14) is surely right when he remarks that
forms of collective provision, like the NHS and
public housing "do represent very important steps
forward and do in part 'enhance welfare'". Thus
"the welfare state is a product of the contradictory
development of capitalist society and in turn it has
generated new contradictions which everyday become
more apparent" (Gough, 1979, 152). These contradictions include the polarisation of the working class
into distinctive, competitive groups or 'fractions'
(see O'Connor, 1973; Westergaard and Rester, 1976).
Also present are the vested interests in the medical
and welfare bureaucracies in which client or patient
interest can conflict with professional definitions
and privileges. It may be that this contradiction
is inherent in any social system in that the necessary administrative apparatus - to determine rights
and entitlements, to set norms and to formulate
policy - develops and takes on interests of its own.
It was of course for such a state of affairs that
Trotsky and Mao Ze Dong advocated permanent revolution.
    A socialist allocational mechanism must, it
seems, always be seen in relation to the goals of a
socialist society. As such, it can be used as a
method of judging administrative arrangements for
enhancing 'equality'. These arrangements must of
necessity involve the use of planning which is seen
as enhancing rather than curtailing individual
freedom (see Laski, 1943; George and Wilding,
1976). But planning is seen as being directed more
at economic than social dimensions of society,
particularly through the socialisation of the means
of production.
    Social planning may be seen as leading to the
humanisation rather than fundamental change of the

existing order (see Miliband, 1969). But concentration on the economic leads to a misshapen view of welfare. As Wilson (1980, 87) comments of earlier struggles: "the class struggle that obtained the Factory Acts and the Beveridge Plan could certainly not be said to have operated in the interest of women who were in both cases defined as home-bound individuals dependent on a male bread-winner." To attempt to ensure that all forms of inequality are dealt with therefore, there must be integrated social and economic planning. And recognition of the fact that centralised decisions and allocation may remove the autonomy of the individual. Campbell's (1978) use of autonomy - the regulation of the self to coexist with the freedom of others - recognises that the social individual must regain consciousness of him/herself as a self-creating member of the human species, find fulfillment in productive labour (or endeavour), develop the higher capacities of human potential and cooperate with his/her fellows. Autonomy is thus produced by 'fraternity', suggesting that all individuals are providers and potential recipients of health care and welfare, leading to the view that the Chinese and Cuban systems of spatially diffuse, low-technology, prevention-oriented mass care are the best examples of socialist allocational mechanisms we have (see Eyles & Woods, 1983).

Such societies have been able to construct their care systems from scratch, although they have built on traditional practices. What of more complex societies? How can a socialist allocational mechanism which paradoxically requires the centralisation of power to ensure equity and its decentralisation to promote fraternity be developed? The answer is of course, with difficulty. But to say that something is difficult is not to say that it is impossible to achieve. We must recognise, along with Taylor-Gooby & Dale (1981), that any strategy to defend and extent welfare services must be part of a programme for developing socialism. But in Britain, the Labour Party's approaches to 'develop socialism' have failed to integrate social and economic planning. Indeed, the Alternative Economics Strategy for the 1983 Election (AES) was a comprehensive economic strategy (see Walker, 1984), although there is no clear agreement about its precise structure (see Rowthorn, 1981). Calls for a social counterpart to the AES also accepted the division between the economic and the social (cf Meacher, 1982). Further, as Hall (1983) points out the AES (and the

1982 and 1983 plans) embodied a public burden model
of welfare with such spending occupying a subsidiary
position to boosting demand and regenerating
industry. Social spending is simply seen as a
problem of financing. It is dependent on economic
growth and "redistributions will occur painlessly,
through growth, and with the restoration of full
employment" (Jordan, 1982). There is a failure to
recognise that socialist restructuring (just as with
the market-oriented restructuring of the late 1970s
and early 1980s) involves painful decisions. But
whereas market decisions do not appear to be human
ones, socialist ones do. But there can be no
gainers without losers as the regional resource
allocation programme has shown (see also Thurow,
1981). The AES also failed to recognise that in its
centralisation, it was elitist and undemocratic,
appealing to particular interests (those represented
by powerful trade unions) rather than a generalised
interest. We should also add that there is little
recognition of the variable requirements of different places: the 'backwaters' will still, it
appears, be cleansed by the strong tides of economic
growth.

The importance of economic growth must not be
minimised but a socialist strategy requires other
elements as well. Clearly, there must be specific
minimum protection and rights (see Pinker, 1979;
Lang, 1979; Townsend, 1979). But minima are
extremely hard to establish, particularly in
societies which value the achievement of maxima.
"If greater equality of whatever kind is desired it
is necessary to reduce economic inequality. To do
this successfully, however, it is necessary to
reduce the hold of the ideology of inequality on
people's values and beliefs ..." (Le Grand, 1982,
150). Education thus plays an important role in
establishing socialist principles of justice which
must contain a consensus over who gets what in the
form of a harmonisation of interests. Mishra (1984)
has interestingly pointed to the cases of Austria
and Sweden. The latter has tried to integrate
social and economic policy to ensure full employment, economic growth, egalitarian ('solidaristic')
wage structure and a high level of welfare and to
minimise social disharmony. Austria is more corpo-
ratist, instituting the idea of social partnership
to ensure the voluntary cooperation of labour and
capital over a wide range of issues. There is a
broad consensus on the objectives of economic
growth, full employment and social protection which

enhance economic and political stability. Both
Sweden and Austria have, it should be noted,
developed a consensus within the confines of a mixed
economy. They possess considerable income inequalities. In practice, however, they differ little to
that advocated by Eurocommunism. And they are
exemplars of what Mishra (1984) calls the integrated
welfare state (IWS).

Fig. 2.2: Types of Welfare State

Differentiated welfare state (DWS)* (Keynes-Beveridge)	Integrated welfare state (IWS)* (Post-Keynesian)
Economy-Regulation of the economy from the demand side. Government measures of 'pump priming', deficit financing, fiscal and monetary policies to stimulate or inhibit demand.	Economy-Regulation of the economy from both demand and supplyside, eg. profits, investment, wage levels, inflation, labour market conditions. Regulation and consensus-building (with or without statutory instruments) across wide ranging economic issues.
Social welfare-Relatively autonomous realm seen as distinct from the economy. State provision of a range of services seen as 'socially' oriented with little explicit linkage with reference to the economy.	Social welfare-Not seen as a realm autonomous of the economy and economic policy. Interdependence and interrelationship between the social and economic recognised and institutionalised. Functional relations and trade-offs between the economic and the social inform policy-making.

Polity-Characterised by interest-group pluralism. A free-for-all or market model of the polity and societal decision-making process. Free collective bargaining in the industrial area. Pursuit of sectional interests through organised groupings, parties and parliament. Exercise of economic power without social responsibility. Parliamentary forms of government. Full civil and political liberties.	Polity-Characterised by centralised pluralism. Bargain between peak associations/representatives of major economic interests over a broad range of economic and social policies. Interdependence of economic groups recognised and institutionalised in the form of class cooperation and social consensus. Major economic power groupings assume social responsibility. Parliamentary forms of government. Full civil and political liberties.

*The terms 'differentiation' and 'integration' are well established in social theory. As used here, the differentiated welfare state refers to the notion of a set of institutions and policies added on to the economy and polity, but seen as a relatively self-contained, delimited area set apart from them. The integrated welfare state suggests that social welfare programmes and policies are seen in relation to the economy and polity and an attempt made to integrate social welfare into the larger society.

Source: Mishra, 1984, 1C2-3

Figure 2.2 portrays the ideal types of the IWS and for comparison the differentiated welfare state (DWS) of Britain. The IWS sees the close inter-relation of social and economic policy, recognising the costs of the social and the need for policy coordination. This coordination is predicated on the view that the cooperation and agreement of major social groupings are necessary for sustaining a productive market economy and a highly developed system of welfare. It must be planned co-ordination but such planning must be voluntary so as to be democratic. We may note that the IWS is still part of a capitalist society. Indeed Mishra, (1984, 1C5) recognises that "corporatism provides an institutional framework for sustaining full employment and comprehensive social services in the context of a liberal market society" (emphasis original). Just

as the DWS wa a step forward from the market economy
so too is the IWS a step along the road to collective responsibility. But "it cannot be a means to
socialism, to an egalitarian society" (Mishra, 1984,
171). The pursuit of 'equality' must take other
forms. Indeed Mishra is, for the time being, silent
on the political feasibility and institution of the
IWS in a society like Britain. But the first thirty
years of the post-second world war era must stand as
testimony to the unlikelihood of its feasibility and
institution unless income redistribution is tackled.
Indeed in his discussions of area deprivation
Townsend (1976) has pointed to the importance of
fiscal and social security policies in removing
social ill-being.

But where does this leave socialist allocational
mechanisms? We must answer again in difficulty. It
is necessary to conceive of structural change in
industrialised nations as occurring over a long
period of time (see Walker, 1984) but it is of
course just as likely that capitalist (market)
principles will re-emerge as for example happened in
the Lucas alternative plan for employment and
investment (Wainwright and Elliot, 1982). How is a
socialist framework instigated without that framework first being in place? Elaborate utopias are
naive and unlikely to accord with everyday life (see
Ferge, 1979). Further, the establishment of the
social priorities of socialist planning (equality of
outcome, decentralised planning, democratic participation, need satisfaction, harmonisation of interest, integration of intitutions) is problematic
without the power to establish the priorities as
concrete practice. It is, however, necessary to
develop a socialist vision to oppose the dominant,
conservative vision of society and to use that
vision to judge the progress of policies towards
socialist ends. The experience of Third World
societies is important (see Ferge, 1979; Conyers,
1982), but the vision must be relevant to and con-
sonant with the lives of people in an industrial
nation. A significant dimension of that vision
involves health as well-being (see Chapter One).
The creation of this vision - this consciousness -
is a major, if not the major, task of socialist
thought.

Conclusion

As it must, this chapter in its discussion of
socialist allocation has gone beyond health as

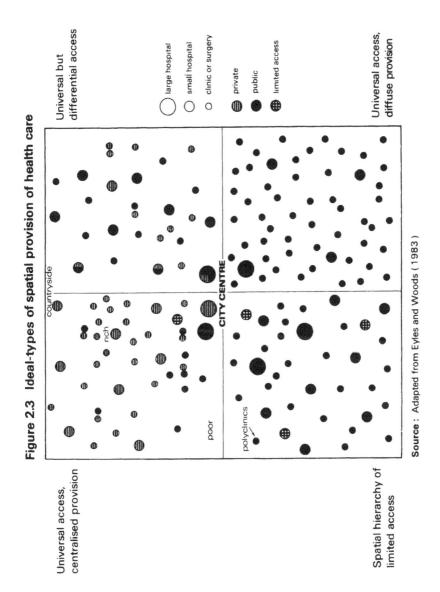

Figure 2.3 Ideal-types of spatial provision of health care

Universal but
differential access

large hospital
small hospital
clinic or surgery

private
public
limited access

Universal access,
centralised provision

Universal access,
diffuse provision

Spatial hierarchy of
limited access

countryside

rich

poor

CITY CENTRE

polyclinics

Source : Adapted from Eyles and Woods (1983)

well-being and the practice of particular societies
to consider again philosophical questions. Just as
Figure 2.1 indicated, all the elements in this
chapter are interrelated. Need, justice and allocation are all intertwined because they are all parts
of the questions of how we allocate resources and
why we allocate them (to others). These parts, as
the discussion on socialist allocation has indicated, are dimensions of societal structure and
values themselves. Thus for example, appeals to
conscience can serve market interests as well as
society as a whole. Indeed Plant et al (1960) note
how 'community' is used by marxists, liberal-
conservatives and interventionists to justify their
particular stances on social welfare (for the
broader theme see Eyles, 1965a). The reason is
simple. Community, like welfare and health as
well-being, is essentially a contested concept.
They are all valued achievements but the means of
getting there are contested (the how). Indeed one
of the purposes of this chapter has been to demonstrate this inherent contestibility. Another purpose as stated above has been to demonstrate the
interrelationships between types of need, types of
justice and types of allocation. Further these
suppose particular spatial outcomes. Figure 2.3
demonstrates the ideal-type of these outcomes for
particular allocational mechanisms, and hence
systems of justice and hence rights, deserts and
needs. These of course refer to the spatial provision of facilities to allocate health care
resources. It is, however, our purpose to examine
the implications of an integrated approach to a
broadly conceived 'health'. Such resource allocations are themselves contested, their integration
being in terms of particular systems of justice and
allocation.. This task will be confronted in Chapter
Five.. Before that we wish to establish further the
case for an integrated approach to health as well-
being first by examining (Chapter Three) the question 'who', with our discussion taking the form of
an analysis of the social bases of health and illness in Britain and secondly, by adding as empirical
dimension to our conceptual elaboration of 'what' in
Chapter One by looking at health and health care
resources in relation to housing, environment, welfare and so on in Chapter Four.

Chapter Three

HEALTH AND ILLNESS IN BRITAIN

Introduction

The principal purpose of this chapter is to describe
the epidemiological context in which health care is
delivered in contemporary Britain, i.e. who is ill
and needs care and why. Traditionally medical
geographers have contributed to descriptive epidemiology through cartographic presentations of, for
example, infant mortality rates, and particular
disease distributions. Howe's (1970) National Atlas
of Disease Mortality is one of the best known works
in this vein, and like the publications of the
Office of Population Censuses and Surveys on area
mortality follows the tradition established by
William Farr in the mid nineteenth century. Subsequently medical geographers have extended their
research endeavours beyond description to explore
relationships between mapped patterns of disease and
other social and demographic characteristics of the
population. Different scale of analysis have been
employed - international, inter-regional, intra-
regional, and intra-urban - to elucidate links
between particular diseases and possible aetiological factors. Numerous studies can be cited (see
Giggs (1979) for a review) as examples of research
in this tradition, which has much in common with the
approach adopted by John Snow in his much quoted
work on cholera in London.
    An alternative approach to cartographic presen-
tation and geographical analysis is to describe the
distribution of disease within populations by reference to some attribute of the individual other than
place of residence. Age and sex are the two attributes commonly used (often in conjunction with the
geographical approach) but more recently attention
has focused on occupation and social class which
have been the subject of intensive investigation

(DHSS, 1980, Townsend and Davidson, 1982). Other
social characteristics of increasing interest are
marital status and race.

At this point it is pertinent to reflect on the
interaction of the two approaches outlined here.
How do social class distributions relate to geographical patterns? If we examined the social class
distribution of a particular disease by, say,
regional health authority would we find the same
class distribution in all regions? Likewise, can we
assume that the health and illness experience of
someone in social class I is the same in South
London as in South Humberside? Evidence from the
Registrar General's decennial supplement on occupational mortality suggests they are not; in other
words geography and social class affect mortality
rates independently of each other (OPCS, 1978). We
develop this theme in greater detail later, but in
attempting a descriptive epidemiology of modern
Britain we wish to emphasise that the classifications we have adopted imparts an unjustified simplicity. In Fig. 3.1 we have attempted to create a
framework which captures some of the relationships
between the categories we use for our descriptive
epidemiology.

We suggest that five principal social characteristics can be used, and three principal geographical
scales. As Fig. 3.1 shows we recognize important
relationships between the social characteristics
themselves (for instance between ethnicity and
class) and with geographical scale. Thus geographical patterns of disease presented using a small
(local) spatial scale (e.g. electoral ward within a
health district) are prone to the effects of concentrations of people with particular social characteristics. As we progress up the geographical scale we
can expect the effect to lessen as population size
increases and the influence of localised concentrations on the overall pattern diminishes. On the
other hand, it is important to remember when working
at the larger scale of regional health authorities
that any calculated rates are based on a denomiator
population of considerable social diversity.

With these points in mind we now proceed to our
epidemiological description of contemporary Britain.
We commence with an overview which describes health
using indicators such as infant mortality and life
expectancy and compare Britain's experience with
that of other nations. We then examine this
generalized picture in greater detail by reference
to the social characteristics of age, sex, occupa-

tion, social class, and race. Finally, we explore
the geographical patterns of illness and disease in
contemporary Britain.

Britain's Health in the 1980s: An Overview

Describing health is a difficult business; so
difficult in fact that it is normally described in
terms of illness and death, the absence of disease
and low mortality rates being generally accepted as
indicators of health (see Chapter One). Measures
such as life expectancy (at birth and other ages),
infant and perinatal mortality rates are amongst the
most commonly used health indicators, and in general
they show two things about contemporary Britain.
First, Britain's population currently enjoys a
better state of health than at any time in the past
(Fig. 3.2) and, second, a state of health better
than most other nations. In Britain life expectancy
at all ages and for both sexes has lengthened over
the past 100 years, dramatically so in the case of
life expectancy at birth. A male child born in 1980
could expect to live for about seventy one years and
baby girls could look forward to seventy seven
years. In 1870 the corresponding life expectancies
were only 40 and 43. Undoubtedly the major factors
which have brought about this change have been the
decline in infant mortality and the successful control of infectious disease.
    Internationally, the health experience of
Britons compares favourably, though it is by no
means the best. The lower levels of mortality
experienced in Britain compared with third world
countries is well recognized but compared with some
other western European nations, Britain has a relatively high infant mortality rate (Table 3.1). In
1980 infant mortality rates ranged from 6.9 per 1000
live births in Sweden to 26.0 in Portugal, the rate
in the UK of 12.1 per 1000 being bettered by eleven
other western European nations. Notwithstanding
this, Britain's infant mortality rate is on the
decline, continuing a trend begun during the last
century. The current rate is now only a third of
the rate recorded in 1948 (O.H.E., 1984). The
decline in perinatal mortality rates has been even
more spectacular. Over the relatively short period
1974-81 the perinatal mortality rate in Scotland
declined by almost 50% and in England by 42%. Even
so, the 1981 rate for the UK as a whole (12.1 per
1000 live and stillbirths) remains relatively high

Table 3.1 Infant Mortality in Western Europe

Country	1980 (except where shown)	
	Deaths under 1 year	Deaths under 28 days
Austria	14.3	14.0 (1975)
Belgium	11.0	9.1 (1977)
Denmark	8.4	5.6
France	10.0	6.0 (1979)
Greece	18.0	14.5 (1979)
Iceland	7.7	6.7 (1975)
Ireland	11.2	6.9
Italy	14.3	11.2 (deaths under 30 days)
Norway	8.1	7.1 (1975)
Netherlands	8.6	5.5
Portugal	26.0 (1979)	15.7 (1979)
FDR	12.6	8.6 (1979)
UK	12.1	8.3 (1979)
Spain	11.1	11.0 (1977)
Sweden	6.9	6.3 (1975)
Switzerland	9.1	7.3 (1975)
Luxembourg	11.5	5.3

Sources: UN (1975) Post-war demographic trends in Europe and the outlook until the year 2000 New York; Eurostat (1982) Demographic Statistics 1980, Luxembourg; Council of Europe (1982) Recent Demographic Developments in the Member States of the Council of Europe, Strasbourg
From Hall & Ogden, 1983, 30

Figure 3.1 A descriptive classification of the National Health

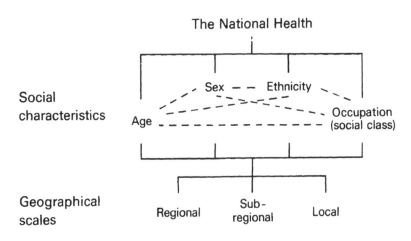

The National Health

Social
characteristics

Geographical
scales

Regional   Sub-
regional   Local

Figure 3.2 Male and female deaths in the UK

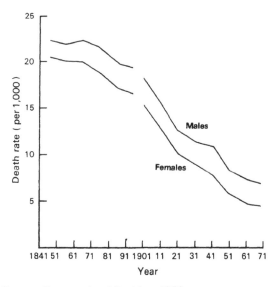

Source: Townsend and Davidson 1982

by the standards of Sweden.

Accompanying the decline of infant and overall mortality has been a change in the causes responsible. In the nineteenth century the most prominent causes were the infectious diseases. Appalling social conditions, poor nutrition and inadequate and ineffective medical care conspired to create an environment in which infectious diseases flourished and people, especially children, perished. Nowadays deaths from cholera and tuberculosis, though not unknown, are certainly uncommon, their decline being attributable to better nutrition, improved housing, clean water and safer food preparation and storage. The principal causes of death are now chronic degenerative diseases of the circulatory system, various types of cancer, respiratory disease and accidents. Fig. 3.3 presents data relating to 1981 showing the number and proportion of deaths in broad disease groupings. Dominating the picture are deaths due to diseases of the circulatory system, ischaemic heart disease (including heart attack) accounting for 54% of the 319,472 deaths in this category. Neoplasms (a category including cancer) accounted for 145,775 deaths of which, by far, lung cancer (27% of deaths in this group) was the most important cause. Over 90,000 deaths were caused by respiratory disease, of which 64% were due to pneumonia. Unlike the infectious diseases which are commmonly caused by single organisms and which result in death after an acute illness, the diseases which currently cause the majority of deaths are of multifactorial aetiology and are often associated with long periods of chronic morbidity prior to death.

Not all disease is fatal, however, and any assessment of health must look beyond mortality to patterns of morbidity in the population. Most morbidity data derive from medical consultations, and to that extent incidence and prevalence rates may be regarded more accurately as measures of the propensity to consult rather than the true level of community morbidity levels. Last (1963) described this phenomenon as the clinical iceberg whereby the illness seen by medical professionals represents only the visible tip of the morbidity iceberg, the greater volume of ice (disease) being submerged from view. Various estimates of size of the hidden disease volume have been made. Cooper (1975, 13) reports for instance that, "for every case of diabetes, rheumatism, or epilepsy known to the GP there appears to be another case undiagnosed. In the case

Figure 3.3 Analysis of mortality in Britain by disease groupings (1981)

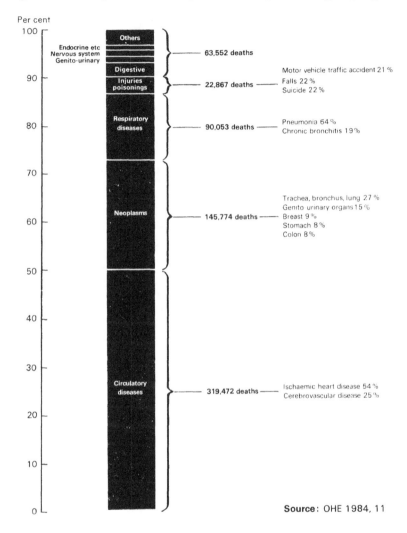

Per cent

100 ── Others

Endocrine etc
Nervous system ── 63,552 deaths
Genito-urinary

90 ── Digestive

Injuries
poisonings ── 22,867 deaths ── Motor vehicle traffic accident 21 %
Falls 22 %
Suicide 22 %

80 ── Respiratory
diseases ── 90,053 deaths ── Pneumonia 64 %
Chronic bronchitis 19 %

70

Neoplasms ── 145,774 deaths ── Trachea, bronchus, lung 27 %
Genito urinary organs 15 %
60 ── Breast 9 %
Stomach 8 %
Colon 8 %

50

40

30

Circulatory
diseases ── 319,472 deaths ── Ischaemic heart disease 54 %
Cerebrovascular disease 25 %

20

10

0 ── **Source:** OHE 1984, 11

Figure 3.4 Days of certified incapacity in Britain by cause (1981-2)

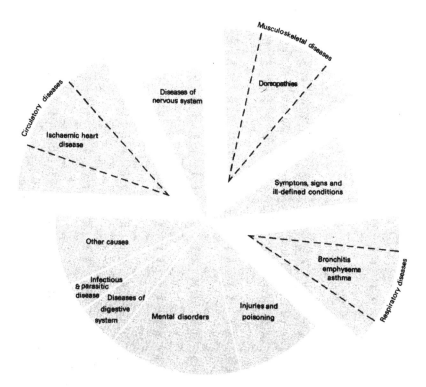

Source: OHE 1984, 19

of psychiatric illness, bronchitis, blood pressure,
glaucoma, and urinary infections, there are likely
to be another five cases undiscovered, whilst the
untreated cases of anaemia probably exceed the
treated eightfold".

Naturally the diseases which are major causes of
death also represent a major proportion of all
morbidity, but there are additionally many diseases
and disabilities which affect many people for long
periods of their lives and which are non fatal.
Psychiatric disorders and handicaps are obvious
examples. It is estimated that each year 5 million
people consult a general practitioner about a mental
health problem (commonly some form of neurosis), and
up to 600,000 receive specialist psychiatric
treatment (DHSS, 1976c). Only respiratory disease
is a more common reason for consulting the general
practitioner (Brearley, 1978). Nearly half a
million people in England are reported to have some
form of mental handicap of which 160,000 are
classified as severe. There are in excess of 3
million people aged over 16 living in private households with some degree of functional handicap
excluding blindness and deafness; over 111,000
people are registered as blind, and up to 4 million
people have significant hearing loss (DHSS, 1984).

One morbidity data set based on interviews which
ask people about their health experience independently of medical consultations is that created by
the General Household Survey (GHS). This is a
continuous survey based on a sample of the population resident in private households in Gt. Britain.
Two principal categories of illness have been
investigated by the GHS; acute (short term illness
in the 14 days prior to interview) and chronic
(longstanding illness experienced all the time or
which keeps recurring). In the 1983 survey 21% of
men and 31% of women over 16 years of age reported
acute illness and in the case of chronic illness the
reported rates were 21% and 25% respectively (OPCS,
1985). It appears then, that at any one time less
than half of the population aged over 16 considers
itself free from illness. These perceptions are
important for they are the starting point for potential patients seeking medical care.

According to the National Morbidity Surveys of
the Royal College of General Practitioners (RCGP)
which are based on returns from over 50 general
practices, the highest consultation rates are for
respiratory disease, mental illness and circulatory
disease (RCGP, 1974). DHSS data on certified incap-

acity - derived from the issue of sick notes by
doctors - confirms this pattern. Fig. 3.4 shows the
proportion of all days of certified incapacity by
diagnostic group in 1981/82. In that year 358.5
million working days were lost due to ill-health
(compared with 4.26 million days lost due to industrial stoppages in 1981 and 5.3 million in 1982) the
principal causes being the familiar circulatory,
respiratory and mental disorders, and in addition
musculo-skeletal disorders (notably back pain). Few
doubt the unpleasantness of illness, but it is also
clear that in the narrowest economic sense it is
extremely costly.

Social Characteristics and Health

Age

Table 3.2   1983:   Chronic and Acute Sickness by Age
            and Sex (percentages)

	Males		Females		Totals	
Age	Chronic	Acute	Chronic	Acute	Chronic	Acute
0-4	11	15	9	14	10	15
5-15	17	12	13	11	15	12
16-44	23	8	23	13	23	11
45-64	44	11	45	15	44	13
65-74	58	12	63	19	61	16
75 & over	67	17	70	20	69	19
Total	31	11	33	14	32	13

Chronic sickness: prevalence of reported limiting
                    long standing illness
Acute sickness: prevalence of reported restricted
                    activity in the 14 days before
                    interview

Source: OCPS (1985), 161, 162

Currently one of the most significant demographic
trends associated with the decline in mortality is
the ageing of the population and the increasing
number of people who have survived a lifetime of
labour and celebrated their sixty-fifth birthday.
Between 1948 and 1982 the population aged between 65
and 74 increased by 39% and those aged over 75 by
94%. Consequently in 1982, 8.3 million people in
the UK were aged over 65 (14.7% of the total populations compared with 5.3 millions (10.75%) in 1948

Figure 3.5 Mortality rates per 1000 population, UK, averages for 1949-50 and 1979-80

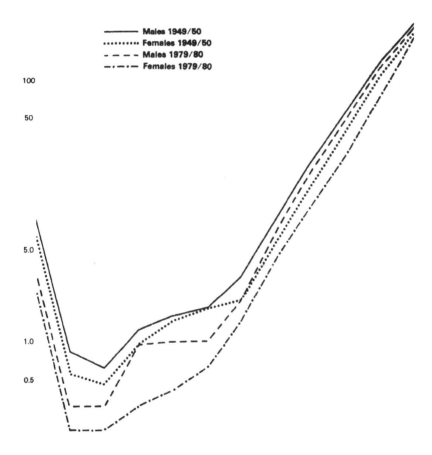

Rate per 1,000
population
500

———— Males 1949/50
··········· Females 1949/50
– — – – Males 1979/80
—·—·— Females 1979/80

100

50

5.0

1.0

0.5

0.1
0-4   5-9   10-14   15-19   20-24   25-34   35-44   45-54   55-64   65-74   75-84 85+
Age group

Source: OHE 1984. 9

Figure 3.6 Distribution of deaths by age-group, percentages and cumulative frequency curves: Britain (1981)

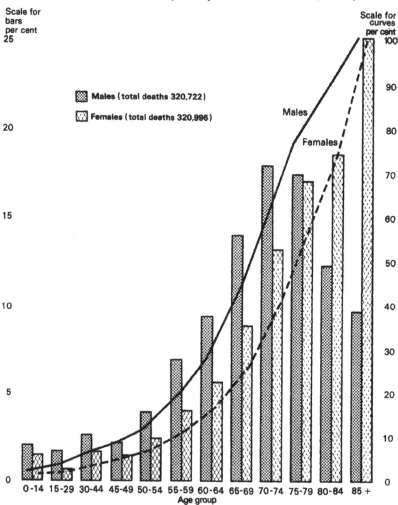

Source: OHE 1984, 10

(OHE, 1984). Growth in the numbers of people in these age groups is of special significance to the health and social·services since it is the elderly who experience the poorest health of any sector of the population reflected in higher morbidity and mortality rates (Figs. 3.5 and 3.6). Mortality rates rise rapidly after the age of 34 and morbidity data from the GHS mirror this pattern (Table 3.2). Thus, 67% of men and 69% of women aged over 75 reported chronic health problems compared with 23% for both sexes in the age group 16-44.

In addition to age differences in the level of morbidity and mortality different causes of illness and death affect different age groups. As Fig. 3.8 shows the principal killers of those under one are perinatal causes and congenital abnormalities. Between the ages of one and fourteen accidents and cancer predominate, as is the case of those aged between 15 and 44. Motor vehicle accidents are a particularly important cause of death in these age groups. After the age of forty five accidents cease to be amongst the principal causes of death. Instead heart disease and cancer predominate, males being particularly prone to lung cancer and females to breast cancer. Deaths amongst those aged over 65 are due to various forms of heart disease, cancer and stroke.

Sex

At various points in the discussion so far differences in the health experience of men and women have been indicated. McKeown's (1979) analysis of temporal change in mortality rates between 1841 and 1971 shows this is not a new phenomenon, men experiencing higher rates of mortality throughout the period than women (Fig. 3.2). The pattern of morbidity, however, does not conform to this relationship. According to the GHS women report more chronic and acute illness, female prevalence rates being higher in all age groups than male rates (Table 3.2).

Roberts (1976, 16) identifies four categories of disease which exhibit sex differences in their incidence. These are, "those conditions primarily or predominantly of a genetic nature; conditions in which environmental factors act unequally on the two sexes; conditions in which environmental factors might be expected to act equally but in which the incidences are unequal in the two sexes; and conditions directly related to the sex organs." For the epidemiologist those conditions in which environmental factors are implicated are of special interest. The first category of "unequal environments" is essentially a function of occupational hazard. Thus lung diseases associated with quarrying and mining are more common in men, whilst housewives, nurses and hairdressers are more likely to suffer from dermatitis as a result of handling soaps, synthetic detergents, cleaning solutions and the like. More difficult to unravel are the sex differences in conditions in which the environmental risk is thought to be equal. Roberts suggests that genetic factors may be partially responsible, but notes that it is not clear how these lead to the higher incidence of arthritic diseases amongst women or the higher incidence of cancer amongst men.

Although more boys are born live than girls the differential mortality experience of the sexes eventually reverses this situation. In the first week of life the chances of a male baby dying are higher than a female, irrespective of gestational age, the overall sex ratio of male, female deaths being 1.25: (Roberts 1976). In childhood male mortality from accidents (the principal cause of death) is strikingly higher but it is in the older age groups that the sex differences in mortality are greatest which is partially a reflection of wartime mortality. However, an analysis of sex differences in mortality and cause of death shows that between 1910 and 1965 the widening of sex differences in mortality in the age group 60-79 (which accounted for 46% of all the sex differences in mortality in 1910 and 63% in 1965) was due to increased male mortality from cardiovascular, renal and neoplasmic diseases. The net result of these factors in the UK is that by the age of 50 women outnumber men, and by the age of 80 the number of women exceed the number of men by about two to one.

Occupation

Given that a man living to the age of 70 will have spent about 60% of those years in (and over half of the days in each year) at work it is reasonable expectation that occupation will have some impact on health. Apart from the immediate dangers of injury undertaking various jobs (and Gray (1979) reports that there are nearly half a million accidents - 1000 of them fatal - at work of each year) the workplace environment itself can be harmful to health. The importance of adequate heating, ventilation and

working facilities has been long recognized and is
reflected in nineteenth century legislation intended
to improve working conditions in particular industries, and more recently in the passing of the Health
and Safety at Work Act 1974.

The range of work related disease is considerable; chemicals are recognized as being especially
dangerous, being reflected in the phrase 'mad as a
hatter' which is attributed to the observed effects
of mercury on felt hat makers (OPCS, 1979). Since
1946 when the National Insurance (Industrial
Injuries) Act became law a number of diseases known
to be caused by particular agents found in the
workplace have been prescribed for the payment of
industrial injury benefit. Amongst diseases on the
list are mesothelioma caused by asbestos; farmer's
lung caused by the inhalation of spores found in
hay; bladder cancer caused by napthylamine; and
pneumconiosis caused by the deposition of dust in
the lungs (Gray 1979).

Establishment of the link between workplace
hazards and particular diseases owes much to epidemiological enquiry which has isolated the harmful
agents responsible. One of the problems in this
kind of enquiry is the length of time between exposure and disease onset; thus in the case of asbest-
osis the effects of exposure only become apparent
many years later. Careful analysis of mortality and
occupation can, however, indicate possible associations. William Farr recognized this as long ago as
1861 when he prepared an analysis of cause of death
by occupation. Nowadays the Registrar-General
prepares an annual statement on occupational mortality, and a decennial supplement which provides the
most comprehensive analysis of the link between
occupation and mortality. Currently the 1970-72
decennial supplement (OPCS 1979) is the most recent
report available; mortality rates are derived by
using a numerator population of deaths by cause and
occupation (as recorded on death certificates issued
in 1970-72) and denominator populations recorded in
the 1971 Census of Population. In the 1970-72
decennial supplement occupations are grouped into
223 occupation units which combine to form 27 occupation orders. This makes it possible, for
instance, to distinguish the mortality experience of
fishermen (occupation unit 001) from gardners and
groundsmen (occupation unit 005) both of which are
in occupation order I (farmers, foresters and
fishermen).

In Table 3.3 an example of the data available

Table 3.3   Standardized Mortality Ratios for Men aged 15-64 by Class and Occupation Group

Class	SMR
I	77
II	81
III (non-manual)	99
IV	106
V	114
Unclassified	137
Unoccupied	100
	36

Occupation Group		SMR	Occupation Group		SMR
I	Farmers, foresters, fishermen	91	XV	Construction wokers	111
II	Miners & quarrymen	144	XVI	painters & decorators	111
III	Gas, coke & chemical makers	107	XVII	Drivers of stationary engines	103
IV	Glass & ceramics makers	109	XVIII	Labourers	141
V	Furnace, forge & foundry workers	122	XIX	Transport & communications	111
VI	Electrical & electronic workers	104	XX	Warehousemen & storekeepers	108
VII	Engineering & allied trades	104	XXI	Clerical workers	99
VIII	Woodworkers	96	XXII	Sales workers	90
IX	Leather workers	114	XXIII	Service, sport & recreation workers	116
X	Textile workers	110	XXIV	Administrators & Managers	73
XI	Clothing workers	103	XXV	Professional workers	75
XII	Food, drink & tobacco workers	110	XXVI	Armed forces	147
XIII	Paper & printing workers	91	XXVII	Inadequately described	86
XIV	Makers of other products	84		All men	100

Source: OPCS 1979, 9

from the decennial supplement is presented.
Standardized mortality ratios (SMR) are used as the
mortality measure, which is the percentage ratio of
observed deaths in each occupational group to those
expected if the death rate of all occupational
groups (the standard population) had applied. In
this case the mortality ratio of men in England and
Wales aged 15-64 is the standard population (SMR =
100) and the mortality experience of occupational
orders and units is expressed as deviations about
this standard. SMRs in excess of 100 indicate an
unfavourable mortality experience, whilst SMRs less
than 100 indicate an unexpectedly low rate of
mortality.

From the table the highest SMR (147) is recorded
by occupation order XXVI (armed forces) and the
lowest (73) by occupation order XXIV (administrators
and managers). An analysis of the constituent
occupation units within the occupation orders
reveals that even larger variations in mortality
experience are recorded. Within order XIX (transport and communication workers) the SMR ranges from
a high of 223 (unit 116, deck and engine room rat-
ings, barge and boatmen) to a low of 81 (unit 129,
postmen and mail sorters).

Extending this analysis to cause specific mortality demonstrates the great utility of the occpa-
tional mortality data, for associationo between
particular causes of death and occupations may
suggest some aetiological factor in the workplace.
It is possible to identify occupations which have
unusually high or low SMRs and this can form the
basis for further enquiry; for instance occupational unit 044 (electroplaters, dip platers and
related workers) records low ratios for all of the
disease groups, except neoplastic disease with an
SMR 58% above the average. In fact, OPCS (1979)
makes some suggested associations between cause of
death and occupation, e.g. accidental falls and
steel erectors (Unit 034), diseases of the
musculoskeletal system and watch and chronometer
makers and repairs (049), tuberculosis of the
respiratory system and miners (101) and kitchen
hands (163), and mental disorders and telegraph and
radio operators (128).

One aspect of occupationally related illness
which now attracts increasing attention is the
psycho-social context of work. Stress and boredom
are two obvious dimensions. Often regarded as a
problem for the high pressured business executive,
occupationally related stress is in fact much more

widespread. Concern over job security, jobs with a high degree of responsibility for others, and no job at all are recognized as stress factors. Boredom, lack of job satisfaction, the constant repetition of tasks with little skill are on the other hand just as likely to lead to frustration, fatigue, and absence from work. By far the largest proportion of certified days of absence from work due to sickness is for short term absence because of minor respiratory and mental symptoms, whereas long term sickness absence (people off work for more than six months) is caused principally by diseases of the heart and circulatory system (1/3) and bone and joint disease (1/5). Overall, sickness absence from work is estimated to cost (through the payment of national insurance, loss of earnings, and lost productivity) the same as the running costs of the NHS (Gray 1979). Quite clearly, many of these costs derive from the hazards of employment itself.

On the other hand, a number of studies have shown that health can improve following unemployment and retirement (see Watkins, 1984) but there is overwhelming evidence that serious health hazards are associated with unemployment. Brenner's (1973, 1979) observations of mortality and unemployment led him to test this relationship on British data. Brenner showed that in the period following increases in unemployment death rates from homicide and suicide also increased, especially within one year. He was also able to show that mortality from other chronic causes increased to three years after the rise in unemployment, Brenner attributing this lagged effect to the degenerative nature of the diseases. Subsequently there has been considerable controversey over Brenner's work. Gravelle et al (1981) were particularly critical of the statistical specification of the multiple regression model used by Brenner, concluding that the model was incorrectly specified. However, Gravelle et al wrote that the absence of a statistically significant relationship between unemployment and mortality increases does not mean that health is unaffected by the impact of unemployment. The loss of income, dignity, self respect and increasing despair in the face of long term unemployment are factors which may have a profound influence on physical, mental and social health. This seems to be accepted by most researchers in this field, (see Farrow 1983 for a review) and has received statistical support from recent studies. Using data from the OPCS Fox and Goldblatt (1982) and Moser et al (1984) examined the

subsequent mortality of the million men included in
the study who were seeking work in the week prior to
the 1971 Census. The results show that by 1981 the
overall death rate for these men was 36% higher than
for employed men of comparable age. Once
standardization for the social class distribution of
the men (see next section) had been made the excess
mortality amongst the unemployed remained at 21%.
Consideration of mortality by cause shows some
corroboration of Brenner's original findings.
Unemployed men were more than twice as likely to
commit suicide in the 10 years after 1971 as
employed men of the same age. Similarly they were
60% more likely to die as the result of an accident,
and 75% more likely to die of lung cancer. Stress
was suggested by Brenner as an explanatory factor,
and to test this Moser et al (1984) examined the
mortality rates of women living with unemployed men,
the implication being that the stress generated by
unemployment is shared by the whole family, and
indeed the findings show that the observed mortality
rate was 20% higher than that to be expected on the
basis of age alone. A comprehensive review of
health and unemployment has recently appeared (see
Smith 1985; 1986) and Moser et al (1986) have
related health, unemployment and area by showing
that the adverse health effects associated with
unemployment are greater in areas of high levels of
unemployment than in areas with low levels.
    Dramatic as this picture is, it may actually
underestimate the true magnitude of the unemployment
effect because they combine both short and long term
unemployed, and because unemployment is corrently
four times higher than it was in 1971. The tensions
which may lead to suicide are more likely to be
heightened in those who have been unemployed for
long periods, and Moser et al (1984) observe that
the differential health experience of these long
term unemployed might be even worse than the figures
he has been able to calculate. Kasl et al (1975)
found that health deteriorated when people lost
their jobs, but recovered when they found new
employment. Stafford, Jackson and Banks (1980)
studied school leavers before they attempted to join
the labour force and afterwards. Those who had not
found jobs had worse health than those who had.
Studies of unemployment and mental health indicate
similar relationships (see Platt, 1984) with, in
general, unexpectedly high rates of psychiatric
illness amongst unemployed people. Although Farrow
(1983) rightly warns of the difficulty in separating

out the precise effect of unemployment in the subsequent deaths, we must concur that whilst the evidence is circumstantial it is also strong. We also concur with his view that "medicine has a dual role. It must respond to the distress of the individual, whether of unemployed teenager or of those made redundant in late middle age. It also has a duty to speak out against those policies that lead to increased unemployment and are likely to be responsible for that distress and for an increased mortality among the unemployed and their relatives." (Farrow, 1983, 105).

Social Class

As well as varying according to occupation there is ample evidence that health experience varies according to social status. For the mid nineteenth century for instance Howe (1972) presents data which illustrates the differential life expectancies of men classified as gentry, tradesmen and labourers, ranging from 50 years to 33 years in Wiltshire and from 35 to 15 in Liverpool. Likewise William Farr compared mortality rates of what he called the professional, the commercial, the domestic, the agricultural and the indefinite and non-productive classes (OPCS, 1979). Interest in analysing mortality data in this way originates from the observation that the lifestyles and resources of the wealthy differ markedly from those of the poor, and that in turn there differences might be expected to have some impact on health independently of occupational factors. Thus, whilst the social classes used in contemporary analyses are derived from groupings of occupations these groupings are "an attempt to rank occupations in terms of skill and to some extent also in terms of standing in the community," (OPCS, 1973, 30). Currently the social classes used by the Registrar General are as follows:

I	Professional (e.g. accountant, lawyer)	5%
II	Intermediate (e.g. manager, schoolteacher)	18%
IIIN	Skilled non-manual (e.g. secretary)	12%
lIIM	Skilled manual (e.g. bus driver, butcher)	38%
IV	Partly skilled (e.g. bus conductor, postman)	18%
V	Unskilled (e.g. cleaner, labourer)	9%

(percentages are of the total number of economically active and retired men).

This classification should not be confused with
the Socio-Economic Groupings (SEG) of occupations
which are used by the General Household Survey to
create its own set of 'social classes' which aims to
create SEGs containing "people whose social, cultural, and recreation standards are similar," (OPCS,
1973). In this discussion social classes derived
from this classification are referred to as GHS
socio-economic groupings.

Leaving aside the question of the adequacy of
the classifications (see OPCS, (1973) and DHSS
(1980) for detailed discussions) they have been used
widely to describe variations in health experience
amongst the British population. The most complete
analysis was prepared by a DHSS Working Group on
Inequalities in Health (known as the Black Report)
published in 1980. Drawing on occupational mortality data and GHS morbidity data the Black Report
demonstrated the magnitude of class inequalities in
health experience, inequalities observable at all
stages in the life cycle. Fig. 3.7 is a graphic
portrayal of the Report's evidence on infant and
child mortality. The consistency of the inverse
relationship between social class and mortality is
striking (see also Table 3.3). What this means is
that the risk of death at birth and in the first
month of life of babies born into a family in social
class V is twice that of a child born into a family
in social class I. In the case of children aged
between one and fourteen the class mortality ratio
(V to I) is, for boys, 2 to 1. Class inequalities
narrow in adult age groups - especially around
retirement age - but in general the higher mortality
rates in the manual working groups persist.

When class gradients are examined by cause the
importance of exposure to environmental factors in
health is clearly demonstrated. The steepest childhood mortality gradients are observed in deaths due
to accidents, the principal cause of death in this
age group. According to Townsend and Davidson
(1982, 53), "boys in social class V have a ten times
greater chance of dying from fire, falls or drowning
than those in social class I. The corresponding
ratio of deaths caused to youthful pedestrians by
motor vehicles is more than 7 to 1." Diseases which
account for the majority of adult deaths - accidents, cancer, heart and respiratory disease - all
exhibit similar relationships in both sexes.

Morbidity data mirror the mortality relationships. Scottish data derived from medical examinations of children aged 5 show that children in

social class V families were, on average, 6 cm
shorter, and a higher proportion suffered from
refractive error in eyesight and tooth decay than
their social class I equivalents. GHS data
pertaining to adults indicate that the relationship
is sustained in later years. The highest reported
rates of chronic ill-health were reported by semi-
and unskilled manual workers (Fig. 3.8) in all age
groups and both sexes (Table 3.4) even though it is
recognized that people in these social classes tend
to under-report their morbidity (Blaxter 1976).

Ethnicity

The use of the word ethnicity creates some problems
of definition. In this discussion we use it to
differentiate the health experience of minority
population groups who share a common cultural heritage. It is not synonymous with race or colour
though these may be important aspects of ethnicity,
nor does it relate solely to immigrants since many
members of Britain's ethnic groups were born in
Britain. The diversity of ethnic groups in Britain
is considerable. Although descendants of families
from Africa, the Caribbean and the Indian Sub-
continent are commonly thought of as the ethnic
minorities our definition includes other groups
amongst which are Greek and Turkish Cypriots,
Vietnamese and Chinese. Racial, religious and other
cultural factors differentiate these groups from
each other and from the native population of
Britain. Referring to these groups collectively as
the "ethnic minorities" is somewhat unfair as it
diminishes this diversity, but all groups do share
two things in common; their experience of discrimination and poverty.
    In the context of health it is possible to
identify at least three separate strands to ethnic
research. Firstly, there has been a concern with
diseases which affect particular ethnic groups
because of some genetic predisposition to the
disease, examples being sickle cell anaemia and
thalassaemia. Secondly, there has been an interest
in diseases which were once common in the indigenous
population, but which are today more prevalent in
some ethnic groups, examples being rickets and
tuberculosis. Thirdly, there is a growing interest
in the sociological aspects of health, that is with
the definitions, meanings and values which ethnic
groups hold with respect to health and illness, and
which seem particularly relevant in the context of
psychiatric illness.

Figure 3.7 Class and mortality in childhood (males and females 0-14)

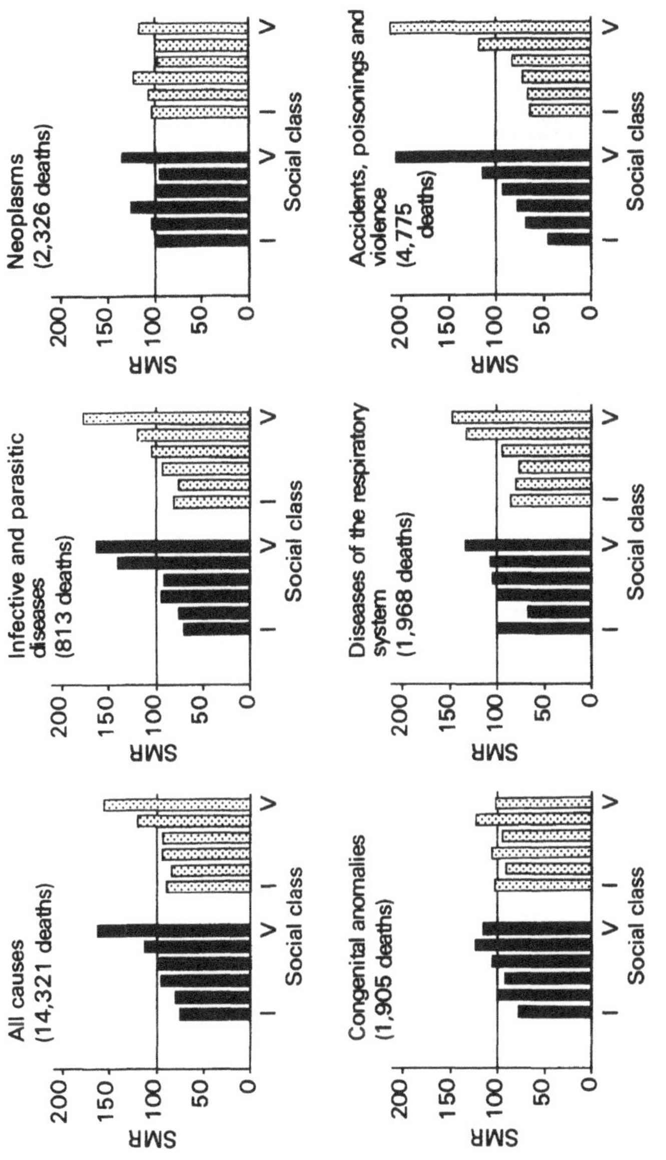

Source: DHSS 1980, p37

Figure 3.8 Morbidity by Socio-economic group, Britain, 1981

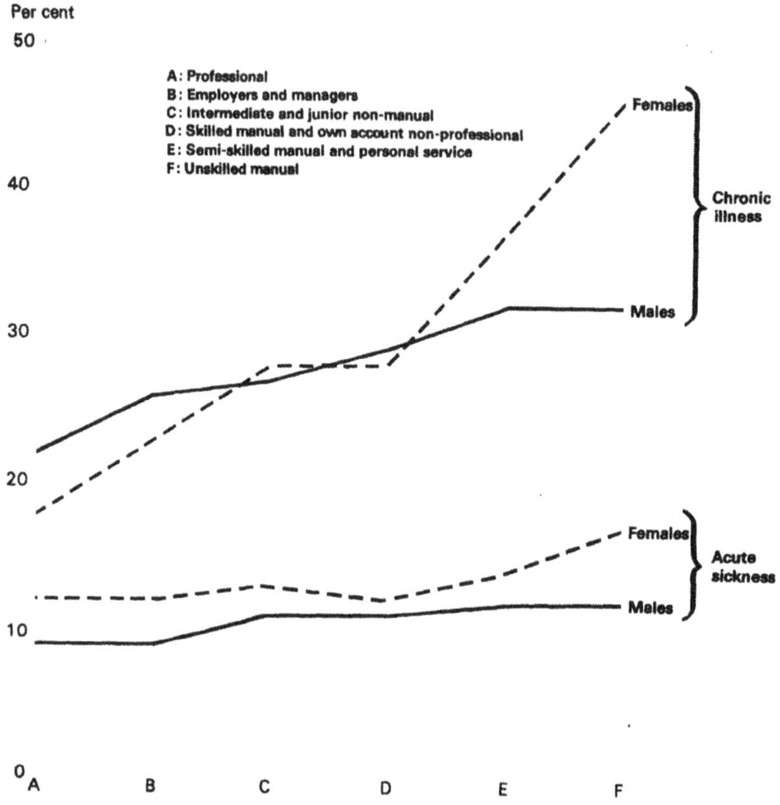

Per cent

50

A: Professional
B: Employers and managers
C: Intermediate and junior non-manual
D: Skilled manual and own account non-professional
E: Semi-skilled manual and personal service
F: Unskilled manual

40

30

20

10

0

A   B   C   D   E   F

Females

Chronic
illness

Males

Females

Acute
sickness

Males

Source: OHE 1984, 15

Table 3.4 Chronic-Ill-Health (Long-Standing Illness) by GHS
Socio Economic Group, 1974-76

GHS S.E.G. Socio-economic group	Age Group					
	15-44		45-64		65+	
	M	F	M	F	M	F
1 Professional	145.4	138.2	228.9	291.3	400.0	376.2
2 Managerial	149.7	141.9	257.0	265.7	476.6	525.6
3 Intermediate	164.0	145.4	368.0	329.7	503.2	553.4
4 Skilled Manual	161.9	167.2	357.7	315.1	541.9	556.4
5 Semi Skilled Manual	173.8	170.3	367.6	380.8	549.8	592.4
6 Unskilled	197.4	202.3	485.5	401.6	542.5	586.2
All	163.2	157.8	348.6	329.4	521.0	564.6
Ratio 6/All	1.21	1.30	1.39	1.22	1.04	1.04

(rates are all per 1000 population)

In all of these categories most research has
been dominated by biomedical perspectives and much
of it has been in fact undertaken by medical professionals focussing on relatively small numbers of
affected individuals (Donovan 1984). On the other
hand interest in the health concerns of the majority
has been limited, even though the focus of such
research are matters manifest in the everyday lives
of all ethnic groups. We might even go so far as to
suggest that ethnic health research has been biased
by the operation of an "inverse interest law" whereby the unusual illnesses of a minority attract the
majority of research interest and the health concerns of the majority are relatively neglected (see
Eyles & Woods (1986) for a fuller discussion of this
'law'). This is not to say that diseases like
sickle-cell anaemia are unimportant. Clearly they
are important; the severity of symptoms and high
mortality rates demand attention, but there is a
danger of concentrating on the unusual, relatively
rare illnesses, converting them into "ethnic
minority problems" and at the same time diverting
attention away from the major determinants of ethnic
health status which remain poverty, environment and
discrimination.
One of the problems confronting ethnic health
research is the absence of data. Ethnic group is
seldom collected routinely in official statistics so
that most information on ethnic disease patterns has
come from special studies. This partly explains the
bias towards the unusual illnesses of rickets,
tuberculosis, sickle cell anaemia, and thalassaemia.
Commonly, official statistics only enable compari-
sons to be made on the basis of place of birth and
this is clearly inadequate in the case of people who
are the second and third generation descendants of
former immigrants. Nonetheless classification on
this basis does reveal some surprising results. In
Table 3.5 SMRs by ethnic group indicate that "men
born in India, Pakistan or the West Indies seem to
live longer than their British-born counterparts"
(Townsend and Davidson, 1982, 59). This favourable
experience is most marked in men in occupational
classes IV and V, the reverse situation applying to
those in classes I and II. Other data produced by
Fox and Goldblatt (1982) confirms this picture
(Table 3.6). Both sources agree that the reason
lies in the self-selection of healthy migrants. As
Townsend and Davidson (1982, 59) observe, "men and
women prepared to cross oceans and continents in
order to seek occupational opportunities or a new

Table 3.5 Mortality (SMR) by country of birth and occupational class (SMR) (males 15-64)

Country of birth	I	II	IIIN	IIIM	IV	V	All
India and Pakistan	122	127	114	105	93	73	98
West Indies	267	163	135	87	71	75	64
Europe (including UK and Eire)	121	109	98	83	81	82	89
UK and Eire (including England and Wales)	118	112	111	118	115	110	114
England and Wales	97	99	99	99	99	100	100
All birth places	100	100	100	100	100	100	100

Source: Townsend & Davidson Inequalities in Health 1982

Table 3.6 Mortality in 1971-75 of males born in selected countries of the New Commonwealth by parents' countries of birth

Place of birth	Parents' countries of birth											
	Both born in the same country as their son			Both born in British Isles or Old Common-wealth			Other combina-tions			All countries of birth		
	Obs	Exp*	SMR	Obs	Exp*	SMR	Obs	Exp*	SMR	Obs	Exp*	SMR
Indian subcontinent	35	52.3	67	19	13.6	140	11	16.8	65	65	82.7	79
Caribbean N Commonwealth	21	25.9	81	3	1.9	158	-	2.7	-	24	30.5	79
Arican N Commonwealth	5	5.4	93	2	1.2	167	2	3.1	65	9	9.7	93

* In this table, expected deaths are calculated using death rates (in five-year age-groups) for all males in the Longitudinal Study.

Source: Fox and Goldblatt (1982, pp 8-13)

way of life do not represent a random cross section
of humanity." The implication is that the effects
of poverty and discrimination take their toll on
succeeding generations. Thomas (1968) and Khogali
(1979) both show, for instance, that the majority of
tuberculosis cases are not brought into the country
by immigrants but are contracted after arrival. The
same is probably true of rickets and osteomalacia.

On the basis of the above data we could therefore expect the elderly members of ethnic groups to
experience better health than their indigenous
equivalents. Evidence from a study conducted in
Birmingham suggests, however, that this is not so.
A household survey of residents in four inner city
electoral wards enabled a comparison of self
reported health status in Asians, Afro-Carribeans
and Europeans over pensionable age. One of the
principal conclusions reached was that "the ethnic
minority elderly are not significantly healthier
than a European group which is markedly older"
(Blakemore, 1982, 71).

Clearly there are many unanswered questions in
the realm of ethnic health, and more research on the
interaction of ethnic groups and their health
beliefs with European biomedical perspectives is
obviously required. Nonetheless, some of the evidence on British ethnic groups, all of the evidence
on black minorities in the USA and Australia, and on
the black majorities in Zimbabwe and South Africa
indicates that when discrimination and poverty
coincide, health suffers. It should come as no
surprise if further research on ethnic health in
Britain fits this pattern, although recent evidence
is again variable (see Brown, 1984; Tuckett, 1985).

Geographical Patterns of Health and Illness

Inter-Regional Patterns

Howe's (1970) National Atlas of Disease Mortality in
the United Kingdom was the first publication to
attempt a comprehensive cartographic analysis of
disease mortality using standardized mortality
ratios. In the atlas patterns of mortality were
displayed for the then local authority administrative units in the UK using mortality data relating
to the years 1954-1958. Howe showed that overall
mortality was highest in the north and west of the
UK and in the central parts of the major cities.
The highest SMR of all was recorded for the City of
Salford (135) and the lowest for the Isle of Ely in
Cambridgeshire.

Since the publication of Howe's Atlas the cartographic portrayal of disease mortality has become
more common and the pattern revealed by Howe is now
familiar. Currently the most comprehensive data set
on regional variations in mortality are the
Registrar General's decennial supplements on area
mortality (OPCS, 1981), the most recent publication
covering the years 1969-73 . Annual publications
giving mortality rates and SMRs for a variety of
administrative units are now produced (OPCS, 1983,
1984) but the data are not mapped, and the denominator population is based on population estimates
rather than census counts as in the decennial supplement. The annual number of deaths in some
diseases categories is small and so regional variations must be treated with some circumspection.
Consequently the 1969-73 area mortality decennial
supplement provides the most reliable picture of
regional variations in death rates.
    Fig. 3.9 shows the pattern of overall mortality
and morbidity from some major causes of death for
four different age groups. An examination of the
pattern of overall mortality confirms the picture
painted by Howe with, in the statistical sense,
significantly higher SMRs in Wales, the north and
west and significantly lower SMRs in East Anglia and
the south and east, a pattern found in practically
every age group. As deaths from circulatory
diseases are a large proportion of all deaths (51%)
the pattern for circulatory diseases is basically
the same as total mortality (Chilvers, 1978;
Chilvers and Adelstein, 1981). In the case of
respiratory disease the north, north west, and west
midlands have significantly higher SMRs in virtually
all age groups, but in Wales only in those aged over
45. Significantly low SMRs for respiratory disease
are recorded for most age groups in the south east
and south west. With respect to neoplasmic disease
(mainly cancer) the pattern is more complex, and
whilst significantly higher SMRs are observed in
some age groups in some regions, noteably Greater
London, the most striking feature is the significantly lower SMRs of the majority of age groups in
the rest of the south east, south west and East
Anglia. The natural histories of individual
diseases within these broad groupings are very
different, so that whilst we have a generalized view
of higher mortality in some regions the geographical
patterns of individual diseases are much more
complex.
    Perinatal and infant mortality rates however,

Figure 3.9 Regional mortality for selected causes of death

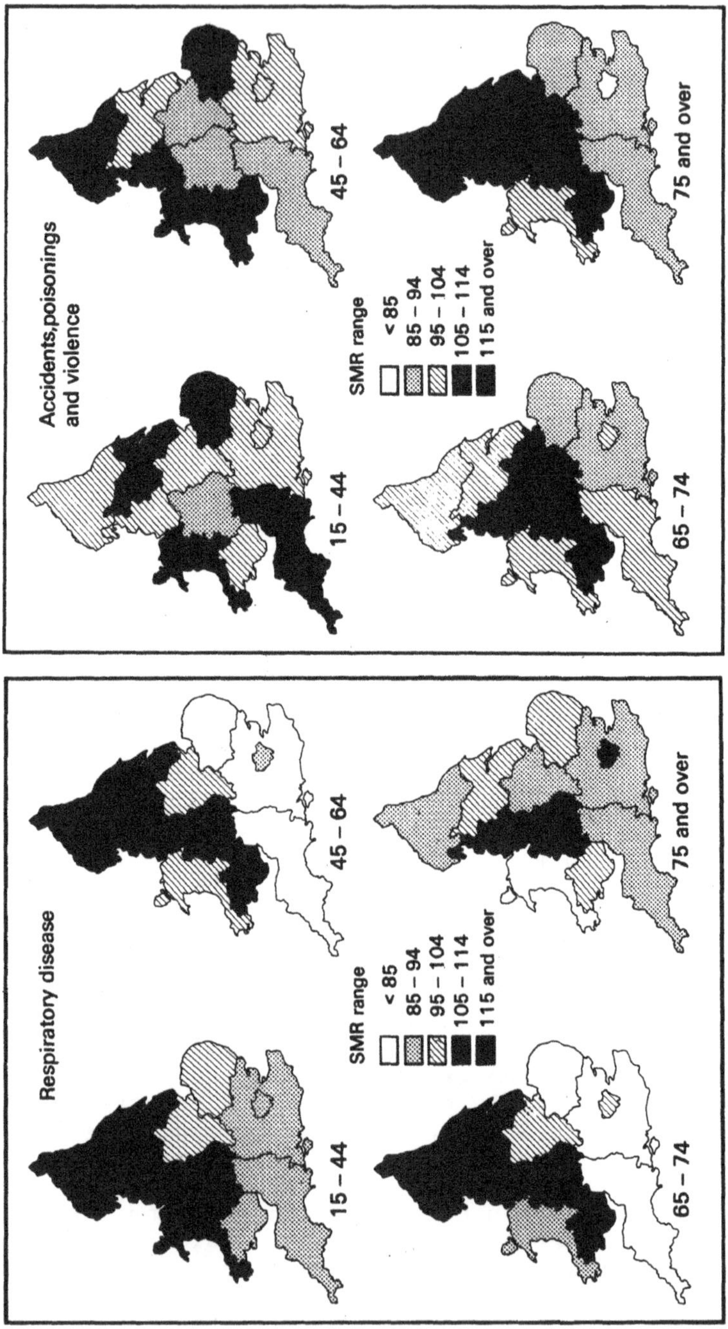

Source: Fox and Goldblatt (1982)

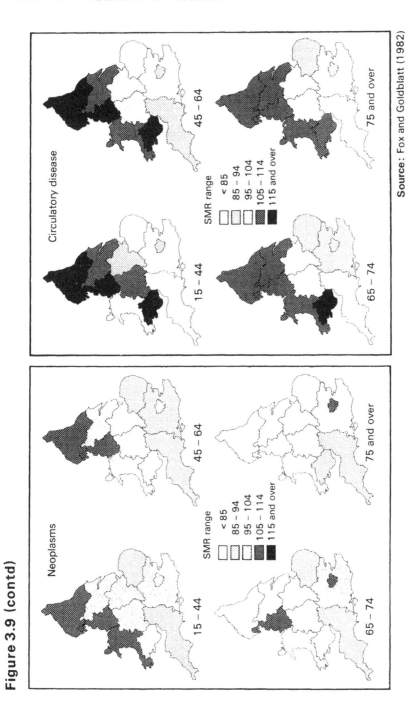

Figure 3.9 (contd)

Source: Fox and Goldblatt (1982)

Table 3.7 Regional comparisons of infant and perinatal mortality

	Infant mortality: rates per 1,000 live births				Perinatal mortality: rates per 1,000 live and still births			
	1974	1976	1980	1981	1974	1976	1980	1981
United Kingdom	16.8	14.5	12.1	11.1	10.6	17.8	13.3	12.0
England	16.2	14.2	12.0	10.9	20.3	17.6	13.3	12.0
Northern	17.3	15.1	12.4	10.7	22.5	19.9	15.0	13.2
Yorkshire	19.4	15.1	12.9	12.6	22.1	18.4	14.9	13.9
Trent	15.6	14.8	10.9	10.9	21.0	19.3	13.0	11.3
E Anglia	14.0	11.7	10.4	9.9	16.6	14.0	11.3	10.4
NW Thames	14.2	13.9	10.9	10.2	17.5	17.0	11.0	10.7
NE Thames	15.7	13.4	12.0	10.4	19.2	16.6	13.8	11.1
SE Thames	15.2	14.8	12.3	11.4	19.7	17.4	12.9	12.1
SW Thames	14.1	12.8	11.0	10.4	18.1	14.9	10.8	10.7
Wessex	14.9	12.7	12.0	11.3	17.8	15.1	11.7	9.9
Oxford	13.6	12.9	11.4	8.2	15.7	13.8	12.6	9.4
S Western	14.6	12.9	11.5	10.4	19.0	16.1	12.6	11.4
W Midlands	16.9	15.7	13.1	11.7	22.5	21.1	15.1	12.8
Mersey	18.6	14.2	12.8	11.3	23.7	19.5	13.7	12.4
N Western	19.8	15.7	12.9	11.1	23.1	18.7	15.3	12.4
Wales	17.0	13.7	11.4	12.6	21.2	19.0	12.8	14.0
Scotland	18.9	14.8	12.1	11.3	23.0	18.3	13.1	11.6
N Ireland	20.8	18.3	13.4	13.2	25.0	22.0	15.6	15.3

Source: Office Health Economics, (OHE) Compendium of Health Statistics (1984) Table 1.3 p 7

Figure 3.10 England and Wales : regional patterns of morbidity

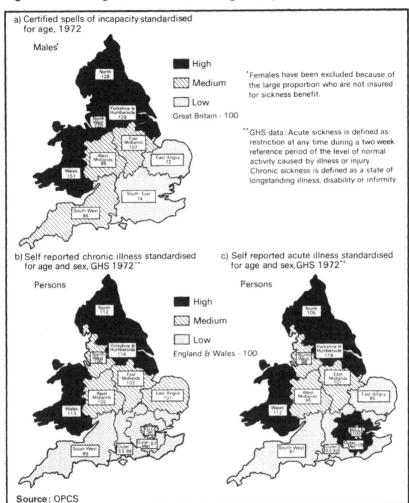

a) Certified spells of incapacity standardised for age, 1972

Males*

High
Medium
Low

Great Britain - 100

*Females have been excluded because of the large proportion who are not insured for sickness benefit.

**GHS data : Acute sickness is defined as restriction at any time during a two week reference period of the level of normal activity caused by illness or injury. Chronic sickness is defined as a state of longstanding illness, disability or infirmity.

b) Self reported chronic illness standardised for age and sex, GHS 1972**

Persons

c) Self reported acute illness standardised for age and sex, GHS 1972**

Persons

High
Medium
Low

England & Wales - 100

Source: OPCS

correspond closely with the overall picture. In
Table 3.7 the infant and perinatal mortality rates
for 1982 of each of the English regional health
authorities is presented along with the rates for
the constituent countries of the UK. Although
England has the lowest infant mortality rate of any
of the home countries (10.9 per 1000 live births)
rates for each of the RHAs varied from a high of
12.6 in Yorkshire to a low of 8.2 in Oxford. The
pattern for perinatal mortality is broadly the same;
Northern Ireland recorded the highest rate (15.3 per
1000 live and stillbirths) compared with lowest rate
of the Oxford RHA of 11.6.

Turning to morbidity data (Fig. 3.10) we can see
that the populations resident in the regions with
the highest overall SMRs also perceive themselves as
experiencing more chronic and acute illness and the
evidence on certified incapacity suggests that
doctors agree with them. Taking all these sources
of data together it is hard to avoid the conclusion
that the healthiest parts of Britain lie south east
of a line from the Wash to the Bristol Channel.

Intra-Regional Patterns: Urban-Rural Differences in Health

The adverse effects of city life upon health have
been recognized for a long time, and since the mid
nineteenth century attempts have been made to modify
the urban environment in the pursuit of health. In
Britain much of the impetus for this movement is due
to the work of Edwin Chadwick who was largely responsible for the preparation for the report of the
Select Committee on the Health of Towns (1840).
This report which ultimately led to the 1874 Public
Health Act and the appointment of medical officers
of health - practising epidemiologists - documented
the health hazards of city life at that time.
Amongst the most important hazards were overcrowding, pollution and inadequate sewage disposal which
created ideal conditions for the proliferation of
infectious diseases. Popular support for the notion
of healthy city life was encouraged in Benjamin Ward
Richardson's book, "Hygeia: A City of Health" in
which he envisaged cities limited in size to a
population of 100,000 with 25 persons per acre, and
houses less than four storeys in height. Medical
services were to be provided in modern hospitals,
special homes created to deal with the problem of
infant mortality, and a principal sanitary officer
appointed to provide leadership for a team of

Table 3.8 Mortality by Age and Sex in Rural Districts and Conurbations 1969-73

		Under 1	1-14	15-44	45-64	65-74	>75
North	M	97	96	109	100	101	104
	F	99	98	111	103	109	108
Tyneside Conurbation	M	106	94	109	122	115	108
	F	111	99	111	113	109	103
Yorks & Humberside	M	95	98	105	89	93	96
	F	103	96	95	95	100	101
West Yorks Conurbation	M	123	116	110	114	112	107
	F	124	114	114	114	113	104
North West	M	96	110	109	99	100	104
	F	93	101	100	100	105	113
SE Lancs Conurbation (Merseyside)	M	123(120)	112(105)	112(109)	119(122)	116(113)	109(106)
	F	123(109)	113(105)	108(105)	117(122)	114(110)	106(104)
East Midlands	M	87	102	99	86	91	97
	F	88	96	95	93	99	98
West Midlands	M	92	115	100	89	92	96
	F	88	128	91	91	99	101
West Midlands Conurbation	M	113	97	99	110	109	105
	F	119	106	101	101	102	101
East Anglia	M	85	94	93	77	79	92
	F	85	82	94	87	90	97
South East	M	72	93	91	80	83	93
	F	78	94	89	86	89	100
Greater London	M	99	103	95	97	98	99
	F	100	103	100	95	93	97
South West	M	83	98	98	86	83	91
	F	90	92	91	90	90	94

Source: Tables 3.4 & 3.5, Area Mortality: Decennial Supplement (1969-73), OPCS, 1981

medical staff, sanitary inspectors and chemists who would deal with sickness in the community. Richardson's city remains the ideal, and though there is no doubt that many improvements in the urban environment have reduced health hazards there are still significant differences in the health experience of urban and rural dwellers. In the most general terms, it appears that the popular image of healthy country life is correct.

Area mortality data from the Registrar General's decennial supplements supports this view. Table 3.8 shows death rate ratios (DRR) for the major conurbations and compares them with the rural districts within the standard regions of England by age group. The death rate ratio (DRR) is the percentage ratio of the age and sex specific death rate to the corresponding England and Wales age and sex specific death rate. Although some rural parts of the regions record DRRs in excess of 100 - especially in the North and North West in people aged over 45, reflecting perhaps the movement of people with chronic work-induced complaints to areas on the urban fringes - they are generally lower in rural areas than in the conurbations. Congestion, pollution, the pace of city life, loneliness, and isolation are amongst the factors responsible for these patterns.

Pollution - particularly atmospheric pollution - is recognized as one of the greatest hazards of the city. Vast quantities of toxic gases are released into the atmosphere by cars, industry and by domestic heat production. Periodically concentrations of pollutants can reach dangerous levels, as happened in December 1952 during one of London's great smogs. During the smog sulphur dioxide levels more than doubled as did the number of deaths registered (Morris, 1975). In the 1950s deaths from bronchitis correlated well with levels of atmoshperic pollution. According to Morris (1975) the male mortality rate from this disease was 76% higher in county boroughs classified as "dirty" than the average for England and Wales, but in rural districts rates were only half of the average. The 1969-73 area mortality data confirm the general relationship between bronchitis mortality and urban areas (Table 3.9). SMRs are, in the statistical sense, significantly higher in conurbations and towns exceeding 100,000 population and significantly lower in towns of less than 100,000 population and in rural districts.

Table 3.9   Urban and Rural Distribution of Mortality
            from Bronchitis, Emphysema and Asthma,
            1969-73

Conburbations	117	122
Urban Population > 1LL,LLL	1L9	1L2
Urban Population 5L,LLL < 99,999	98	97
Urban Population < 5L,LLL	96	89
Rural Districts	76	77
England and Wales	1LL	1LL

Source: Table 4.4L.1, Area Mortality (demand
         supplement) OPCS (1981)

     Although the clean air legislation of the 195Ls
has made an enormous contribution to the diminution
of atmospheric pollution in towns, other pollutants
have become the focus of concern. Radiation and
atmospheric lead are perceived as major threats;
the dangers of exposure to radiation are well known,
but it is only recently that levels of atmospheric
lead have come to the public's attention. High
levels of atmospheric lead produced by the exhaust
emissions of motor cars are believed to be harmful
to young children as extended exposure to high
levels may cause brain damage. Children residing
close to major road junctions are obviously at
greatest risk (Giggs, 1979). In the light of a
government working party report (Lawther, 1961) the
recommended petrol lead level has been reduced from
L.4 grammes/litre to L.15 grammes/litre.
     In addition to the production of health hazards
in urban areas the urban residential environment is
itself a potential health hazard. Overcrowding is
traditionally associated with poor health since it
predisposes the rapid transmission and diffusion of
infectious disease, but it also produces tensions
and stresses in family and social life. Likewise
poor quality housing lacking amenities such as a
bath or a lavatory hinders the adoption of high
standards of personal hygiene, and damp rooms are
know to exacerbate respiratory disease. Housing
design itself may also be a factor in the creation
of stress and strain which manifests itself in
psychiatric illness. High density, high rise living
has been said to create 'tower block neurosis'
(Allen, 1973) as a result of the isolation which
high rise developments impose. Even rehousing
schemes which move families to new towns in rural
settings may simply transfer old problems to new
settings revealing themselves not as 'tower block

neurosis' but as 'new town blues'. In short, towns in terms of their housing designs, estate layouts, and traffic flows represent health hazards in addition to the concentration in them of harmful agents produced by modern industrial processes (see Eyles & Woods (1983) for a review). This is not to see that rural areas are without their own particular health hazards.

Intra-Urban Patterns

Although we have been able to observe that the residents of south east England appear to enjoy better health and longer lives an examination of intra-urban patterns indicates that the health experience of some residents in parts of London is at least as bad as anywhere in the country. In short the regional scale of analysis belies the diversity of health patterns within the city. Donaldson (1976) illustrated the magnitude of these differences in the relatively small conurbation of Teeside which had in 1971 a population of about 400,000. Distinguishing between the inner part of the conurbation ('downtown') and other parts of the city on the basis of housing quality Donaldson examined differences between the two areas by reference to the five characteristics of social and environmental conditions, fertility, mortality, child health and general health. In Table 1.10 the results of this analysis as presented and the higher incidence of social disadvantage and poor health is clearly revealed in the figures for the downtown area.

In the larger contexts of London and Manchester two recent studies have investigated intra-urban differences in morbidity in considerable depth (Leavey, 1982 and Curtis, 1983). Both studies set out to measure the prevalence of morbidity in representative samples of the adult population in two geographically distinct parts of the cities concerned. In the London example the boroughs of Tower Hamlets (inner city) and Redbridge (suburban) were selected and in Manchester the inner city partnership area and the remainder of the Manchester, Salford and Trafford local authority districts were chosen. Both studies employed a questionnaire incorporating a measure known as the Nottingham Health Profile (NHP) developed in the Department of Community Medicine at Nottingham University, and fully described in a series of papers by the research team (Hunt and McKeown, 1980, Hunt et al,

Table 3.10 Urban and Suburban Differentials in Teesside

Indicator		Downtown	Rest of Teesside
% 1971 Resident Population of Teesside		18	82
% Population aged >65 (1971)		12	9.8
% Population aged 5-14 (1971)		16.6	19.7
% in Social Class IV+V (1966)		42	29
% of population born in New Commonwealth		3	0.5
% unemployment (1971)		9.2	5.6
% of house with >1.5 persons per room (1971)		2.6	1.3
% of households lacking exclusive use of inside WC		60.9	6.5
Birth Rate per 1000 women aged 19-40		168.4	79.2
% of illegitimate births		15	7.4
SMR All Causes (1971)	M	137	98
	F	139	90
Ischaemic Heart Disease	M	97	101
	F	125	93
Bronchitis	M	200	74
	F	213	72
Infant Mortality Rate per 1000 live births (1971)		32.4	16.7
% Non Smokers		38	49
Notified TB cases per 1000 residents		0.62	0.27

Source: Donaldson (1976)

1980, Hunt et al, 1981). Its use in the London
study is described in Curtis (1983) and Curtis and
Woods (1984) and in the Manchester study in Leavey
(1982). The NHP is a self completion measure
designed for use with large survey populations.
Part 1 of the profile comprises 38 questions used to
measure six dimensions of morbidity which comprise
the profile (energy level, pain, emotional reactions, sleep, social relationships and physical
mobility). Individual questions are answered yes or
no, replies are weighted and combined to give a
score on each morbidity dimension ranging between 0
and 100. The questions and the weightings were
determined in a series of tests which established
the validity of the profile for various groups of
respondents and in association with clinical
criterian. The individual dimensions of the profile
cannot be aggregated to give one overall score which
reflects health status, but the mean score of survey
groups on each dimension can be calculated for
comparative purposes. It is this approach which has
been used in the London and Manchester studies.
Results from these studies indicate that the profile
is a test of severe morbidity, minor ailments not
being picked up by the questions.
     Data gathered by the profile can be used in one
of two fairly simple ways. Either the mean score of
each group on each dimension can be compared, the
higher score reflecting the presence, on average, of
more severe morbidity in the group, or the proportion of people in the group who registered any score
on a particular dimension can be presented. In this
latter case the severity of individual responses is
ignored, the proportion of respondents registering a
score indicating the prevalence of any morbidity on
that dimension in each group. Both approaches are
presented in the discussion which follows.
     The mean average score on each dimension for
Tower Hamlets and Redbridge and for Inner and Outer
Manchester are shown in Table 3.11. On every dimension the mean score in Tower Hamlets exceeded that
in Redbridge (with the exception of the sleep dimension for the elderly) when the samples were disaggregated by age group. In the case of the
Manchester data there is a similar pattern of scor-
ing. Indeed, not only is the relationship in the
same direction (higher scores in the inner city
sample) but the magnitude of the scores on each
dimension is very similar to those recorded for
London.
     Turning to the proportion of the samples who

Table 3.11 Self-reported Morbidity in London and
            Manchester

Morbidity Dimension	Mean NHP Score for respondents in age group in			
	Inner London	Outer London	Inner Manchester	Outer Manchester
Energy				
16-44 yrs	8.6	7.6	12.4	8.3
45-64 yrs	20.8	10.2	19.4	15.6
65 yrs +	31.8	22.2	32.9	26.7
all aged 16+	15.9	11.1	17.9	14.5
Pain				
16-44 yrs	3.6	1.3	3.2	2.8
45-64 yrs	12.6	3.4	8.0	6.4
64 yrs +	14.8	14.0	16.8	12.9
all aged 16+	8.0	4.3	6.9	6.1
Emotional Reactions				
16-44 yrs	11.4	5.0	11.6	7.6
45-64 years	15.9	8.5	14.2	9.6
65 yrs +	17.5	10.5	16.8	12.5
all aged 16+	13.7	7.1	13.2	9.2
Sleep				
16-44 yrs	11.5	8.9	11.8	7.3
45-64 yrs	20.6	14.8	19.3	14.6
65 yrs +	30.0	32.3	26.6	27.8
all aged 16+	17.6	15.1	16.5	14.0
Social Isolation				
16-44 yrs	5.5	1.7	5.0	3.0
54-64 yrs	10.8	5.5	7.1	5.3
65 yrs +	15.2	13.4	11.2	9.4
all aged 16+	9.6	5.1	6.7	5.1
Physical Mobility				
16-44 yrs	2.2	0.9	2.1	1.7
45-64 yrs	9.9	4.0	6.2	5.0
65 yrs +	23.7	20.4	20.5	16.8
all aged 16+	7.8	5.5	6.6	5.9

Sources: Curtis (1983), Leavey (1982)

Table 3.12 London: Self-reported morbidity by age
                     and area

Morbidity Dimension	Age Group	Proportion scoring more than zero (reporting some morbidity) on the dimension in			
		Tower hamlets		Redbridge	
		%	(N)	%	(N)
Energy	16+	29	(369)	22	(465)
	16-44	17	(201)	15	(229)
	45-64	38	(109)	21	(146)
	65+	51	(59)	46	(85)
Pain	16+	22	(365)	14	(465)
	16-44	12	(201)	6	(229)
	45-64	32	(108)	14	(146)
	65+	39	(56)	39	(85)
Emotions	16+	46	(362)	34	(462)
	16-44	43	(199)	28	(228)
	45-64	36	(107)	46	(145)
	65+	55	(56)	50	(84)
Sleep	16+	46	(367)	42	(462)
	16-44	34	(201)	30	(227)
	45-64	51	(108)	43	(146)
	65+	78	(58)	70	(84)
Isolation	16+	23	(367)	14	(465)
	16-44	15	(201)	6	(229)
	45-64	30	(108)	15	(146)
	65+	35	(58)	34	(84)
Mobility	16+	25	(366)	19	(463)
	16-44	11	(201)	5	(228)
	45-64	30	(108)	21	(145)
	65+	65	(57)	54	(85)

Source: Curtis (1983)

reported any morbidity at all (Table 3.12) we can observe that on every dimension and in every age group (except emotional reactions, age group 45-64) the greatest proportion was found in the inner city sample. In passing we should also note that the proportion of both samples who reported morbidity increases with age. When the NHP data are disaggregated to their individual items (that is the 38 questions) we can observe that the proportion reporting a particular problem was higher in the inner city sample (Table 3.13). Further disaggregation into age groups showed that the proportion of the inner city sample reporting morbidity on any of the 36 items exceeded the outer city proportion in every age group, except for 5 items in those aged over 65 years.

From these data it is clear that the NHP identifies a greater prevalence of morbidity amongst representative samples of inner city residents. The importance of these data is that they are derived from population surveys and not based on doctor-patient consultations which are unlikely to reflect the true prevalence of morbidity in the community because of variations in service availability and because of differences in illness behaviour. Moreover as the NHP is a self completion questionnaire it is not as susceptible to interviewer bias as some other survey instruments. We can conclude with some confidence, then, that illness is more prevalent amonst inner city residents.

Generally explanations offered for the greater prevalence of illness in the inner city highlight the corresponding prevalence of poor quality, over-crowded housing, immigrant populations, social disorganization and other poverty indicators. These are environments which are especially hazardous to health. In the home, lack of space, damp walls, shared or absent amenities, make life increasingly difficult especially for parents with new born infants. In such circumstances it can be difficult to adopt good feeding and hygiene practices. A study of children born in 1979-80 to parents resident in Tower Hamlets and Hackney in East London (Cullinan and Treuherz, 1981, 1982) showed that mothers who feed their babies solely with artificial feeds at four weeks of age tended to be younger and poorer, and that in the first year of life bottle fed babies were more likely to end up in hospital with gastro-enteritis. The risk of admission for upper respiratory tract infections was higher in large, poor families, and especially so in those

Table 3.13 London: self-reported Morbidity by
            Individual Item and Area
(Response to items; all age groups)
% of respondents reporting that item did apply to
them

NHP ITEM (Abbreviated Description)	Tower Hamlets	Redbridge
1 Tired all the time	18	12
2 Pain at night	11	6
3 Things are getting me down	26	13
4 Unbearable pain	5	2
5 Tablets to sleep	18	12
6 Forgotten what it's like to enjoy myself	14	7
7 Feeling on edge	23	14
8 Painful to change position	8	5
9 Feel lonely	13	9
10 Can only walk indoors	6	4
11 Hard to bend	11	9
12 Everything is an effort	12	7
13 Wake in the early hours	39	33
14 Unable to walk at all	0.5	0.4
15 Hard to contact people	9	7
16 Days seem to drag	15	14
17 Trouble with stairs and steps	16	10
18 Hard to reach for things	9	8
19 In pain when I walk	12	7
20 Loose my temper easily	23	16
21 Nobody I'm close to	10	4
22 Awake most of the night	12	7
23 Feel I'm loosing control	6	2
24 In pain when standing	9	5
25 Hard to dress	5	3
26 Soon run out of energy	19	15
27 Hard to stand for long	18	14
28 In constant pain	7	3
29 Takes a long time to get to sleep	21	19
30 Feel a burden to people	5	2
31 Worry keeps me awake	12	4
32 Life is not worth living	4	2
33 I sleep badly	19	15
34 Find it hard to get on with people	7	3
35 Need help to walk outside	6	2
36 In pain climbing stairs/ steps	11	7
37 I wake up depressed	12	5
38 In pain when sitting	8	3

households where parents smoked more than 20 cigar-
ettes per day. Of the babies admitted to hospital
following an accident, "the single outstanding
characteristic associated ... was their mother's
unhappiness with their housing conditions; over
half were either homeless, squatting or trapped in
high rise flats or poor damp houses," (Cullinan and
Treuherz, 1962, 30).

Residents in these areas are not unaware of the
problems, but the resolution of them commonly lies
beyond their control or their means. A survey
carried out in Spitalfields within Tower Hamlets
illustrates his point (Lauglo, 1984). Respondents
were asked to name the health problems faced by
residents in the area, an area representing the
worst of inner city problems. A majority of those
surveyed (57% of all respondents) had no hesitation
in answering 'housing', and 'rubbish disposal'.
Poor quality and overcrowded housing, lack of play
space and uncollected rubbish were ranked higher as
the health problems of the area than "ill health"
per se. Whilst most of these data relate to conditions in the city there are inner parts of the outer
city which increasingly show the same characteristics. Fringe council estates built to accommodate
overspill families from the bombed and congested
central areas are prime examples. As the families
which moved out have aged their children find themselves remote from employment opportunities in
housing and environments which show the signs of
wear. Health levels and problems in these parts of
the city are underresearched but as with their inner
city equivalents the link between environment and
health is not simply an issue of academic interest
for these people; it is a fact of their everyday
lives.

Conclusion

In this chapter we have attempted to look at the
national health from a variety of perspectives, but
as we warned in the introduction there is a danger
in creating a false simplicity when disaggregating
populations by age, sex, place of residence and so
on. In reality the health experience of individuals
is related to an amalgam of the classificatory
variables we have employed. Moreover, individual
health experience is certainly dynamic, changing as
each of the factors we have used exerts a more or
less prominent impact on individuals throughout
their lives. Who is needy and needs care is

Table 3.14 Rankings of Area Health Authorities on Mortality Indices After Standardisation for Social Factors

		1	2	3	4	5	6	7	8	9	10	11	12	13	14
							Mortality Indices								
Area After standardisation															
	Walsall	5	4	6	6	6	3	6	6	6	5	6	4	4	62
	Bolton .	4	4	4	4	6	1	6	6	6	4	6	5	6	58
	Sandwell	6	6	5	4	6	5	6	6	3	6	5	1	5	56
	Wolverhampton	6	3	5	6	6	6	3	6	2	1	4	6	6	54
Worst	Lancashire	2	4	5	6	5	3	2	6	4	4	6	4	4	53
ten	Warwickshire	6	5	4	2	5	4	4	4	4	6	3	6	5	52
	Cleveland	5	1	6	6	6	4	3	3	6	5	6	3	3	52
	Staffordshire	6	4	6	2	4	2	5	5	5	3	6	4	5	51
	Birmingham	5	6	2	5	6	3	4	5	4	2	5	4	4	50
	Bradford	4	5	5	5	4	6	1	5	.2	3	2	6	6	50
	Hampshire	1	5	4	2	3	4	1	4	3	2	3	2	2	35
	Cumbria	3	2	2	5	1	4	4	1	3	1	1	5	4	33
	Suffolk	1	2	2	2	1	2	1	2	5	4	6	3	3	33
	Avon	2	4	1	2	2	4	1	5	1	2	3	1	6	32
Best	Bromley	3	3	4	1	5	1	3	2	2	5	1	2	2	31
ten	Newcastle upon Tyne	3	1	4	3	2	1	4	1	2	4	2	3	4	31
	Gloucestershire	1	2	3	5	4	2	1	4	1	2	3	2	1	30
	Sheffield	3	1	1	1	1	1	2	6	4	6	4	1	1	29
	Oxfordshire	1	1	1	2	2	2	3	3	2	3	4	3	2	28
	Northern Tyneside	4	1	6	1	1	4	1	4	1	1	2	1	3	26

The numbers indicate the sextile group to which each AHA belongs for each disease: 6 = highest mortality group, 1 = lowest mortality group. These scores are then summed (excluding perinatal mortality) to yield an overall rank score and AHAs are presented in decreasing order of this score.

Source: Charlton et al (1983, 694)

Key to Mortality Indices

1	Perinatal Mortality	11	Abdominal Hernia
2	Hypertensive Disease	12	Maternal Deaths
3	Cancer Cervix Uteri	13	Anaemia
4	Pneumonia and Bronchitis	14	Overall Rank Score
5	Tuberculosis		
6	Asthma		
7	Chronic Rheumatic Heart Disease		
8	Acute Respiratory Infection		
9	Bacterial Infection		
10	Hodgkin"s disease		

certainly a complex issue. The intervening factor
in these processes is the availability of medical
care, both preventative and curative. Services are
provided in the belief that health can be maintained
or restored, either through the avoidance of risk or
by interventions in established disease processes.
As with the other variables we have used to classify
health, the availability of medical care is also
differentially distributed. All 'wheres' are not
the same. In a health service which sets out to
provide equal opportunity of access to medical care
for people and populations at equal risk it is
dismaying to find that the NHS seems not to achieve
this especially in those conditions where medical
therapy is known to be effective.

Silman and Evans (1981) have investigated cancer
survival rates for the regions of England and Wales.
They showed that "the more lethal cancers had the
largest relative differences in survival with a
four-fold difference between the best and the worst
regions. Conversely, the less lethal cancers had
the largest absolute differences of up to 25%"
(Silman and Evans, 1981, 291). Variations in cancer
registration practice make it difficult to be precise about the origin of these differences; are
they artefacts of the measurement process or are
they real differences which indicate more effective
detection, diagnosis and management of disease?
Only better data may enable an answer to be found,
but one study has revealed spatial variations in
excess mortality from diseases known to be amenable
to medical intervention once other factors affecting
mortality (as described above) have been taken into
account. Charlton et al (1983) identified fourteen
causes of death which can be avoided through medical
intervention, and then adjusted SMRs to take account
of social factors. The resultant mortality rates
enabled Charlton and his colleagues to rank the 98
area health authorities according to their performance in avoiding mortality from the various causes.
In Table 3.14 the rankings of the ten 'best' and ten
'worst' areas on each of the individual causes are
tabulated, and an overall ranking is presented in
which the areas are distributed in descending order
of summed individual disease rankings. Thus, overall Walsall AHA records the highest level of avoidable mortality, compared with Northern Tyneside
which records the lowest level. As the authors
conclude, "the extent of the unexplained variation
in mortality warrants further investigation, since
mortality from all of these diseases should be

largely avoidable by efficient and effective health
care ... If it is established that there are indeed
large variations in the quality of health care
delivery from one part of the country to another
this will have implications for resource allocation" (Charlton et al 1963, 696). As we shall see in
Chapter 5 there is evidence to indicate that this is
so. And indeed resource allocation policies may not
always improve the situation, with funding being
reduced in places with some of the worst health
problems e.g. inner London. But before we explore
this in detail· we wish to widen the context by
exploring further the links between the epidemiological description which forms the context and other
aspects of the national health which are implicated
in its production. This chapter has identified the
'who', the next extends the definition of the
'what'. 'Health' is not simply a matter of not
being ill. It is well-being (Chapter One) and hence
'health care' implies consideration of housing,
income and so on. If well-being is 'health',
deprivation may well indicate its absence.

Chapter Four

HEALTH AND DEPRIVATION

The account of health and illness in Britain given
in the preceding chapter can be described as descriptive epidemiology, although in our discussion of
patterns of mortality and morbidity we referred to
relationships with work, social and environmental
factors - relationships which are part and parcel of
everyday lives. In this chapter we broaden our
descriptive epidemiology and examine more explicitly
some of these related factors. The 'who' is likely to
be ill where (Chapter Three) is joined with some of
the issues presented in Chapter One. Thus we are
effectively converting our discussion of health into
a discussion of well-being whereby we see health as
one dimension of life experience, bound-up in it
both positively, as a contributor to well-being and,
negatively, as an outcome when health is damaged
through the absence of circumstances which promote
well-being or the others which create ill-being. The
examples from Spitalfields in London's East End
which we quoted in the last chapter underlines the
recognition which people give to the inter-related
nature of health, well-being and material circumstances. One of the respondents in Cornwell's
(1984) study of health and illness in Bethnal Green
articulates these relationships particularly clearly
when referring to the impact of her childhood slum
environment.

> So consequently you was playing with rub-
> bish and filth all the time. You just
> couldn't get away from it all the time. I
> know when I was under the Children's
> Hospital for a year, because I don't know
> whether it was just me, you know, you get
> these kids that are always ill, or whether,
> as I say, it was where we lived. 'Cause it
> doesn't matter how clean you were when you

go to bed of a night, and you don't know
who's in the toilet, because anyone could
go in the toilet because the toilet was
there for anyone. It was like a public
toilet come to think of it (Cornwell,
1984, 32).

In a policy context the most significant recent
statement on these relationships was the Black
Report (DHSS, 1980). Starting from an examination
of health outcomes – some of which we have described
– the policy recommendations of the Report are
distinguished by their emphasis on changes in social
policy to derive health benefits rather than on
health care policy per se. As such, they follow in
a tradition established by the nineteenth century
public health movement and, sustained through the
concerns of Beveridge, and the creation of the post-
war welfare state. The main features of the
report's recommendations are now well documented
(Townsend and Davidson, 1982) and in summary
amounted to an intent to increase the real incomes
of the poorest and disadvantaged by means of state
organized transfer payments. Underlying these
proposals is the belief that the presence or
absence of poverty (however it originates) remains
the major determinant of social well-being, poor
health being but one dimension. Social and geographical inequalities in well-being are sustained in
the form of relative deprivation whereby increases
in real income are unequally distributed according
to the allocative principles of a competitive
society (see Chapter Two). If we are to produce a
geography of the national health and well-being we
need therefore to understand the geography of
poverty. That is the task of this chapter. In
effect, it extends our discussion of both 'who' and
'what'. Those most likely to be ill are the
deprived and they are deprived not just with respect
to health care but income, housing, environment and
so on. Herein lies the basis for an integrated
approach to health as well-being.

Amongst the best known discussions of poverty
are those of Seebohnn Rowntree whose work in York
sought to identify the extent of poverty in that
city (Rowntree, 1901). In Rowntree's survey a
family was considered to be in poverty if its
earnings were insufficient to maintain what Rowntree
termed 'physical efficiency'. The level of earnings
required to avoid poverty was determined by identifying 'minimum necessaries' of food, rent, clothing
and calculating monetary values for these. When

income fell below this level a family was deemed to
be in poverty, although Rowntree distinguished
between 'primary poverty' caused by a lack of income
and 'secondary poverty' caused by expenditure on
items not directly contributing to the maintenance
of physical efficiency. Once incomes exceeded the
level determined by this process families ceased to
be in primary poverty. In time, provided the
incomes of those in primary poverty increased beyond
the established standard, poverty could be eradi-
cated. Thus, Fieghen et al (1977) reached the
conclusion that on the basis of a 1971 absolute
standard the number of people in poverty declined
from 20% of the population in 1953/4 to about 2.5%
by 1973, due not to any income redistribution but to
overall economic growth. The appeal of absolute
measures has led to their widespread adoption in
many different societies, but their use has been
questioned for failing to take account of the social
context in which the poor live. Townsend (1954) has
argued, for instance, that when "considering the
spending habits of poor people..... due regard must
be paid to the conventions sanctioning membership of
their community, to the influence of economic and
social measures currently adopted by society as a
whole..... and to the standards encouraged by the
press, the BBC and the church." This view of
poverty as relative deprivation widens the concept
of poverty beyond the issue of the income required
to maintain a subsistence level of living to encom-
pass the distribution of all forms of resources
available to a society. This distinction is
developed in Townsend's (1979) comprehensive analysis of poverty in the United Kingdom. Three
measures of poverty are employed by Townsend; the
state's standard of poverty based on supplementary
benefit rates; the relative income standard, an
income point which allows a determined proportion of
the population with the lowest income to be identified; and the deprivation standard, a level of
income below which the experience of deprivation
increases disproportionately to income, deprivation
being a reflection of current styles of living in
the society under consideration. It is this last
measure which Townsend develops to reflect his use
of the concept of relative deprivation. In determining income levels which permit participation in
widely accepted styles of living, "it is hypo-
thesised that, at a particular point for different
types of family, a significantly large number of
families reduce more than proportionately their

participation in the community's style of living.
They drop out or are excluded. These income points
can be identified as a poverty line" (Townsend,
1979, 249). A list of sixty indicators of style of
living was established by Townsend, covering diet,
clothing, fuel and light, home amenities, housing,
environment about the home, welfare benefits, family
support, recreation, health and social relations.
From the sixty indicators a summary index of deprivation was derived and is presented in Table 4.1
which shows the proportion of population suffering
deprivation according to each indicator. By allocating a score of 1 or 0 (deprived or not) to each
component of the index Townsend calculated deprivation scores for various families and incomes. The
mean score for adults was 3.5 (maximum 10); a score
of 5 or 6 was regarded as indicating the presence of
deprivation with 20% of households exceeding this
score. After examining the relationships between
mean scores, household sizes and incomes Townsend
reached the conclusion that threshold incomes can
be identified below which the increase in deprivation is disproportionate to the fall in income,
these thresholds being at higher income levels than
those qualifying for supplementary benefits. Table
4.2 compares the percentage in poverty using the
three measures of poverty described above, and it
can be seen that only 6.1% of the population were in
poverty according to the supplementary benefit
standard, compared with 22.9% according to the
deprivation standard. The basic supplementary
benefit level of income used was an extremely severe
test of poverty because it ignores discretionary
benefits, income and assets disregarded in the
determination of benefits. Estimated to be worth
40% of basic supplementary benefit rates, this extra
income was added to the basic benefit rate to identify a 'real' government standard below which people
were considered to be either in or on the margins of
poverty, a total of 15.2 million people - 28% of the
population. In his surveys Townsend found important
differences in the incidence of poverty. Households
consisting of one person, or with five or more
people had the highest rates of poverty, women were
more likely to be in poverty than men, as were
separated and divorced people, and those with
unskilled manual occupations. In particular the
incidence of poverty was higher in older age groups,
especially pensioners living alone, and also in
families with three or more children.
      In the geographical literature a different

Table 4.1 The deprivation index

Characteristics	% of population	Correlation coefficient (Pearson) (net disposable household income last year)
1 Has not had a week's holiday from home in last 12 months	53.6	0.1892 S = 0.001
2 **Adults only.** Has not had relative or friend to the home for a meal or snack in the last 4 weeks.	33.4	0.0493 S = 0.001
3 **Adults only.** Hs not been out in the last 4 weeks to a relative or friend for a meal or snack.	45.1	0.0515 S = 0.001
4 **Children only (under 15).** Has not had a friend to play in the last 4 weeks.	36.3	0.0643 S = 0.020
5 **Children only.** Did not have party on last birthday.	56.6	0.0660 S = 0.016
6 Has not had an afternoon or evening out for entertainment in the last two weeks.	47.0	0.1088 S = 0.001
7 Does not have fresh meat (including meals out) as many as four days a week	19.3	0.1821 S = 0.001
8 Has gone through one or more days in the past fortnight without a cooked meal.	7.0	0.0684 S = 0.001
9 Has not had a cooked breakfast most days of the week.	67.3	0.0559 S = 0.001
10 Household does not have a refrigerator.	45.1	0.2419 S = 0.001
11 Household does not usually have a Sunday joint (3 in 4 times).	25.9	0.1734 S = 0.001

12 Household does not have
   sole use of four amenities
   indoors (flush WC; sink or
   washbasin and cold-water
   tap; fixed bath or shower;
   and gas or electric cooker).   21.4   0.1671 S = 0.001

Source: Townsend (1979)

Table 4.2 The Extent of Poverty

Poverty standard	Percentage of house-holds	Percentage of popula-tion	Estimated number (UK)	
			Households	Non-institu-tionalized population
Rate's standard (SB):				
in poverty	7.1	6.1	1.34 mil.	3.32 mil.
in margins of poverty	23.8	21.8	4.50 mil.	11.86 mil.
Relative income standard:				
in poverty	10.6	9.2	2.00 mil.	5.0 mil.
in margins of poverty	29.5	29.6	5.58 mil.	16.10 mil.
Deprivation standard:				
in poverty	25.2	22.9	4.76 mil.	12.46 mil.
Total (UK)	100	100	16.90 mil.	54.5 mil.

Source: Townsend (1979)

approach to poverty has developed, summarized as a
concern for spatial inequalities in the level-of-
living. Instead of utilizing survey methods to
determine income levels indicative of deprivation
thresholds the emphasis has been on the use of
readily available social indicators which are held
to measure social well-being, level-of-living or
quality of life. Frequently indicators have been
combined to create a numerical index but no attempt
is made to relate levels of living to income
equivalents. The indicators are allowed to speak
for themselves as normative measures of the social
condition. Interest in the development of social
indicators can be traced to the 1960s and a concern
about the reliance on economic indicators as
measures of social progress. The most difficult
problem has always been the selection of indicators
to measure the concept, some studies involving
multi-factorial analysis of census data (Moser and
Scott 1961, 1965; Knox 1974) others searching text
books on social problems (Smith 1973) and others
selecting indicators without any apparent a priori
reasoning apart from their availability (Knox 1982).
In addition to the problem of selection is the
problem of weighting indicators, especially if they
are to be combined into a single index. If indicators are combined unweighted the implicit consumption is that they are all equally weighted, and as
such equally important measures of social well-
being. This might be so; equally it might not, but
it is exceedingly difficult to establish a basis for
allocating weights. Despite these methodological
difficulties which have been fully discussed else-
where (Smith 1973, 1977, 1979; Thunhurst, 1985)
there is no doubt that our knowledge of lifestyles,
opportunities, and handicaps has been enhanced by
the development of social indicators and there is
now a wealth of material relating to geographical
territories of varying sizes.

The Regional Pattern of Deprivation

The origins of contemporary poverty patterns of
regional inequalities in social and economic circumstances are commonly attributed to the inter-war
years which saw the beginnings of a massive restructuring of industry, bringing with it mass unemployment, squalor and suffering on an appalling scale.
But, as the heavy industries of the northern coalfields and shipbuilding towns collapsed, some parts
of Britain benefitted from a new prosperity associa-

ted with the growth of new manufacturing industries,
located in the main, within the south east. Images
of Jarrow marchers, poverty and its main cause mass
unemployment have been etched in the mind, but in
some places unemployment never increased above 5% of
the workforce (Knox, 1982). The same is true today.
Fig. 4.1 compares the pattern of unemployment in
1931 and 1984. The correspondence is obvious, despite the efforts of successive postwar governments
to promote economic development (until the 1979
Conservative victory) outside the south east.

The stability of relative regional prosperity is
captured in Fig. 4.2 in which the "level of living"
of local authority areas is compared in 1951 and
1971 (Knox 1982). The level of living score
employed in this analysis is an unweighted index of
four indicators - overcrowding, male unemployment,
ill-equipped housing, and infant mortality. Large
positive scores indicate deprived places; large
negative scores favoured places. Places lying
within the bold diagonal lines showed no significant
change in their level of living relative to other
places over the period, and 3/4 of places fall into
this category. Of the remainder Knox suggests that
five types of place can be identified based on the
direction of change; places which are 'deprived
divergent', that is places where deprivation has
worsened; 'inverted to deprived', that is places
favoured in 1951 but which had become deprived by
1971; 'favoured convergent', that is favoured places
which have become more so; 'inverted favoured', that
is deprived places in 1951 but which were favoured
by 1971; and 'deprived convergent' places which
still deprived in 1971 but significantly less so
than in 1951.

The picture which emerges from the mass of
information now available indicates the existence of
a prosperous south east compared with the rest of
the UK. Townsend's (1979) poverty survey confirmed
this view, with the caveat that the variation was
less than expected. Nonetheless, as Table 4.3
shows, Northern Ireland had nearly twice as many
people in poverty or on its margins (44.3%) than did
Greater London (23.10%). Likewise, the proportion
of people in Greater London with incomes twice as
large as the supplementary benefit level was 48.7%
compared with 26.4% in Northern Ireland. Generally,
the proportion of regional populations in poverty
increases the further north and west one is from
Greater London. It is worth remembering, though,
that these are measures of the relative incidence of

Figure 4.1 Unemployment rates in the UK 1931 and October 1983

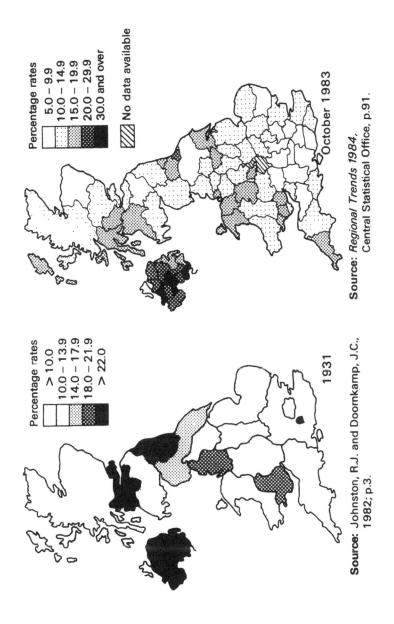

Percentage rates

☐ 5.0 – 9.9
▦ 10.0 – 14.9
▦ 15.0 – 19.9
▦ 20.0 – 29.9
■ 30.0 and over

▨ No data available

October 1983

Source: *Regional Trends 1984*,
Central Statistical Office, p.91.

Percentage rates

☐ > 10.0
☐ 10.0 – 13.9
▦ 14.0 – 17.9
▦ 18.0 – 21.9
■ > 22.0

1931

Source: Johnston, R.J. and Doornkamp, J.C.,
1982; p.3.

Figure 4.2 UK: changes in levels of living, 1951-71

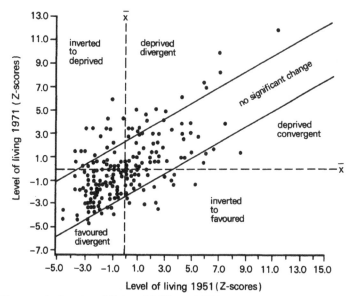

Change in levels of living 1951 to 1971, by local authority
Note : areas experiencing a significant change in level of
living were as follows : *Deprived Divergent:* Armagh, Belfast,
Birkenhead, Bradford, Dewsbury, Fermanagh, Glasgow,
Liverpool, Londonderry, Newcastle, Rochdale, St Helens,
Tyrone, Warrington ; *Inverted to Deprived :* Barnsley, Blackburn,
Bolton, Burnley, Darlington, Grimsby, Huddersfield, Leicester,
Lincoln, Manchester, Nottingham, Oldham, Preston, Salford;
Deprived Convergent : Anglesey, Caithness, Shetland,
Sutherland; *Favoured Divergent :* E.Sussex, E.Yorkshire,
Gloucestershire, Solihull, Somerset, Westmorland, W.Sussex;
Inverted to Favoured : Aberdeenshire, Antrim, Argyll, Flintshire,
Inner London, Kincardine, Montgomery, Nairn, Norfolk, Orkney,
Radnor
Source: Knox (1982)

poverty; in absolute numbers Townsend's estimates
indicated that there were more people in poverty in
London (570,000) than in Northern Ireland (460,000).

**Table 4.3 Percentages of Population in Different
Regions, According to Net Disposable
Household Income in Preceding Year**

Region	Household Income as a % of Supplementary benefit scales		
	Under 140	140– 199	Over 20
	(in poverty or on its margins)		
Northern Ireland	44.3	29.3	26.4
North West	33.9	31.6	34.6
South West and Wales	29.2	38.2	32.7
Scotland	29.1	31.2	39.7
North Yorkshire and Humberside	28.5	35.6	35.9
West Midlands	25.4	33.6	41.9
Anglia and East Midlands	24.9	33.6	41.5
South East	24.2	36.8	39.1
Greater London	23.1	28.3	48.7
All Regions	27.8	28.3	48.7

Source: Townsend (1979, 264)

There is no doubt, though, that the least prosperous regions of the UK lie north and west of line
from the Bristol Channel to the Wash (see Chapter
Three for the similar health situation). Cornwall
is the exception to the south of this line. Data on
the percentage of children receiving free school
meals show this clearly (Fig 4.3). In 1979/80 free

Figure 4.3 UK : receipt of free school meals

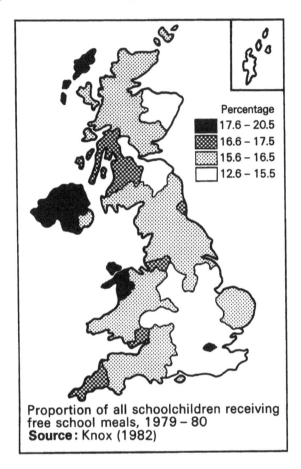

Proportion of all schoolchildren receiving
free school meals, 1979 – 80
Source: Knox (1982)

school meals were a means-tested benefit available
in all local authorities, and as such reflected the
regional distribution of low income. Data from the
Inland Revenue show that the South East had the
largest proportion of tax units (that is tax liable
individuals and married couples) with the largest
incomes. Conversely, Wales, Northern Ireland and
the North West had the highest proportion in the
lowest income bracket. The sources of income show
some interesting variations. In Northern Ireland
less than 60% of households derive their income
from wages and salaries, which is the smallest
proportion of any place in the UK. On the other
hand, it has the largest proportion of households
dependent on social security benefits. As expected
the proportion of households deriving their income
from social security benefits is smallest in the
South East, but the proportion of households
deriving income from pensions is highest in the
South West. The ownership of consumer goods like
tumble-driers, dishwashers, refrigerators,
deepfreezers, televisions and telephones is con-
sistently higher in the south east. In regions
where the proportion of low income households is
greatest, the proportion of weekly income spent on
fuel, food, light and power, clothing and footwear
is highest; where incomes are higher the proportion
spent on these items is smaller, but the proportion
of expenditure on housing and transport is higher.
With respect to spending on tobacco and alcohol, the
proportion of weekly income expended is highest in
the low income regions, the highest actual expenditure on these commodities being in Scotland and in
the North and North West respectively.

The Intra-Regional Pattern of Deprivation

The north-south dichotomy we have described under-
states the complexity of geographical variations in
standards of living. Within the most affluent
regions zones of intense poverty can be found.
Within London, for instance, are districts where
some of the poorest people in Britain live. Conversely, within the poorest regions are places
distinguished by their apparent affluence. But it
is the persistence of the "inner city problem" which
must dominate any considerations of sub-regional
patterns of deprivation. Poverty in the central
parts of cities, or adjacent to them, is not new;
Charles Booth's surveys of East London in the late
nineteenth century documented the plight of East

Enders living next to the commercial centre of the
British Empire. Of a population of 456,878 in 1877
Booth calculated that 35% of residents in the district now known as Tower Hamlets lived on or below
the poverty line including 17%, "for whom decent
life is not imaginable" (Fishman, 1979, 42).   In
this, Walter Besant's City of Dreadful Monotony',
poverty was the norm. As we shall see, for many
contemporary residents, it remains so.
     Over the past two decades interest in the inner
city problems has mushroomed, spurred on by an
upsurge in the number of studies undertaken to give
statistical precision to the description of unequal
living standards within cities. Frequently these
studies have been undertaken to guide social policy,
in the sense that they have been used to target
limited resources on disadvantaged groups resident
in comparatively small areas. Strangely, whilst
this approach was used for housing and education
during the late sixties and seventies, it was not
adopted in the case of health services until
recently. Nearly all of the studies which have
sought to identify zones of deprivation rely on
census data, employing a wide variety of variables
and statistical methods to define and describe
deprivation (see for example, the review by Gitts,
1976, the work of Holterman, 1975, and in the health
care field the work of Scott-Samuel 1977 and
Carstairs, 1981). As we have noted, there are
numerous technical and conceptual difficulties with
these approaches (see Herbert and Smith, 1979 and
Thunhurst, 1985 for reviews of these). It is not
our intention to describe the mechanics of numerous
studies which by applying the "arithmetic of woe"
(Craig & Driver, 1972) have documented the pattern
of deprivation (variously defined). Rather, we are
interested in some of the policy implications which
have flowed from them, and from some important
omissions in the understanding of patterns of deprivation.
     From a policy perspective studies which high-
light pockets of deprivation have some attraction.
They suggest that the problem of poverty is limited
to particular groups of people in particular places,
whose experience of poverty can be relieved by the
allocations of additional resources, or by the
fostering of greater co-ordination between existing
resources and agencies. These area-based positive
discrimination policies are especially appealing if
resources are scarce as they have been since these
policies gained favour in the seventies. Subsequent

research has, however, revealed the limitation of
these approaches (Eyles, 1979, 1986; Hamnett, 1979,
Edwards and Batley, 1978, Jones, 1979, Hall, 1981).
When Barnes (1975) evaluated the Educational
Priority Area (EPA) programme adopted by the Inner
London Education Authority (ILEA) it was found that
for every two disadvantaged children in schools in
EPAs, there were five other disadvantaged children
who were in other schools not in EPAs. After
reviewing the evidence of his own surveys, Townsend
(1979, 56C) concluded that,

> However we came to define economically or
> socially deprived areas, unless we include
> nearly half the areas in the country there
> will be more poor persons or poor children
> living outside them than in them. There is
> a second conclusion. Within all, or nearly
> all, deprived priority areas there will be
> more persons who are not deprived than
> there are deprived. Therefore, discrimina-
> tion based on ecology will miss out more of
> the poor or deprived than it   will
> include.

In spite of this warning, there has been a
recent surge in health-related research which has
sought to identify parts of the city with populations in greater need of health care. As in other
spheres of social policy the majority of such
studies rely on census data to identify what, in the
health care literature, are known as high need
areas. There is a real danger that this line of
reasoning will not heed the lessons learnt in other
social policy arenas and will adopt areal patterns
as a basis for policy. A good example is the work
of Jarman (1983, 1984). Following his membership of
Professor Acheson's London Health Planning
Consortium's Primary Health Care Study Group
(Acheson, 1981) Jarman set out to identify 'under
privileged areas' in which the need for primary
health care was inflated. he did this by asking
general practitioners to identify factors which
according to them inflated their workloads. Then
census indicators were identified, standardized and
combined to form a single index, individual weights
for each indicator being determined through the
survey of GPs (Table 4.4). Using these data, it is
a relatively straightforward task to compute the
value of the index for any census spatial unit or
health authority and this is precisely what Jarman
did. Fig 4.4 presents the results for four

Table 4.4   Average Scores of Social Factors for
General Practitioners in Different Parts
of United Kingdom

		Average Scores		
		Urban England	Rural England	Rural Wales, Scotland, & Northern Ireland
1	Over 65s	6.23	6.20	6.06
2	Under 5s	4.63	4.35	4.84
3	Unemployment	3.32	3.23	3.88
4	Poor housing	4.22	3.36	3.54
5	Ethnic groups	3.39	1.75	1.38
6	Lone parent families	3.31	3.11	2.74
7	Elderly alone	6.59	6.64	6.67
8	Overcrowding	3.37	2.83	2.78
9	Lower social classes	3.86	3.74	3.74
10	Mobility	3.06	2.74	2.51
11	Fewer married families	2.99	2.85	2.36
12	Crime rate	2.88	1.97	2.36
13	Visiting difficulties	3.18	3.60	3.34

Source: Jarman (1983)

different parts of the UK. Examining London in
greater detail, Jarman concluded that the highest
index scores - the most underprivileged areas - were
in Inner London, a finding replicable in most
British cities. Although at pains to point out that
the study was not prescriptive in nature there can
be no doubt that the logical development of the
approach entails the amendment of GP policy with
respect to underprivileged areas, even if the
magnitude and nature of the required change in
policy is unclear. Presumably this is why the BMA
General Medical Services Committee established its
Underprivileged Areas Sub-Committee and funded
Professor Jarman's work.

Subsequently, there have been critiques and
developments arising from Jarman's analyses. Scott-
Samuel (1984) produced an alternative index for the
Merseyside RHA, and Thunhurst (1985) has produced a
map of Sheffield identifying "priority areas of
deprivation" based on a clustering analysis supplemented by a "grassroots survey" to check how the
statistical analysis compared with the views of
"1500 grassroots workers throughout the city, from

teachers, social workers, councillors to the police"
(Thunhurst, 1985, 103-104). As a result 24 priority
areas were expanded to 30, suggesting that the
census indicators fail to capture the concept of
deprivation as it is understood by the grassroots
workers (Fig 4.5).

Clustering techniques have proved popular with
other health care researchers venturing into the
deprivation distribution debate. Morgan (1983) and
Morgan and Chinn (1983) have described the use of "A
Classification of Residential Neighbourhoods" or as
it is better known ACORN Based on clustered census
variables ACORN describes 11 types of residential
neighbourhoods. Morgan and Chinn (1983) examined
variations in rates of morbidity, mortality and
health service use by ACORN, neighbourhood type to
determine if ACORN could be used as an alternative
to social class as a classificatory device. Their
principal conclusion was that ACORN appeared to
differentiate health outcomes at least as well as
social class. But there is a danger. ACORN is
undoubtedly useful as a descriptive measure, but
there is a problem in using it to summarize an
individual's social characteristics - the well known
ecological fallacy. In Morgan and Chinn's (1983)
study ACORN group was allocated to individuals
according to their place of residence (post code);
there is, of course, no guarantee that individuals
possess the social characteristics summarized by the
ACORN cluster.

Ecological analyses are largely responsible for
one of the major shortcomings of area-based
approaches which have been described, and the myth
exposed by Townsend (1979), namely that deprivation
is solely an inner city problem. The high incidences of poverty in inner city areas (with declining
population numbers) distract attention from the
majority of deprived people who live outside such
areas in places with comparatively lower incidences
of poverty but much larger (and increasing) populations. Residents of working class public housing
estates on the periphery of cities and new towns are
examples. Though characterized as places of relative prosperity the outer-city and indeed rural
areas contain numerous people living in poverty. As
the recession of the 1980s has intensified the
number of people affected has increased so the issue
is now clearer than ever before. To quote
Herrington (1964, 24-25)," the closure of firms has
become much more widespread, and even the more
advanced technological industries, which find favour

Figure 4.4 Underprivileged areas : composite scores for family practitioners committee areas

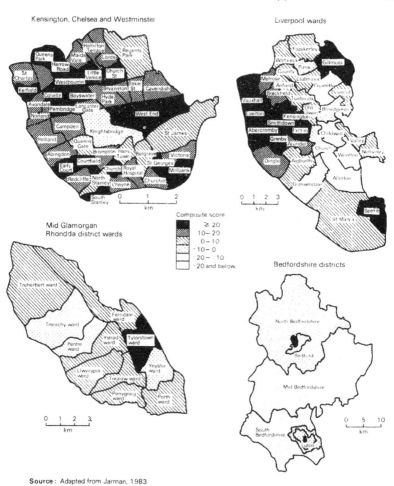

Kensington, Chelsea and Westminster

Liverpool wards

Composite score

≥ 20
10 – 20
0 – 10
–10 – 0
–20 – –10
–20 and below

Mid Glamorgan
Rhondda district wards

Bedfordshire districts

Source : Adapted from Jarman, 1983

140

Figure 4.5 Areas of deprivation and poverty

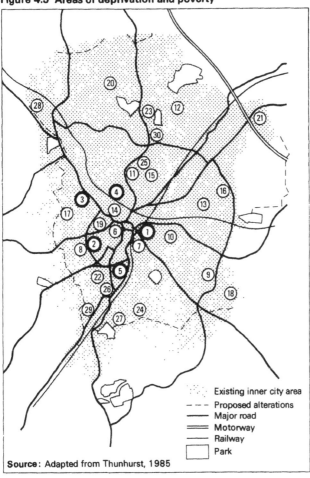

Source: Adapted from Thunhurst, 1985

Existing inner city area
Proposed alterations
Major road
Motorway
Railway
Park

AREAS OF ACUTE POVERTY

1   Hyde Park
2   Broomhall Flats
3   Kelvin
4   Pye Bank
5   Cromford Street

AREAS OF POVERTY

6   Crofts Buildings/Townhead/
     Hawley Street
7   Park Hill
8   Havelock
9   Manor
10   Wybourn
11   Nottingham Street
12   Flower Estate

13   Staniforth Road
14   Moorfields Flats/
     Gibralter Buildings
15   Ellesmere
16   Greenland-Darnall
17   Crookesmoor
18   Woodthorpe
19   Edward Street
20   Parsons Cross
21   Tinsley
22   Sharrow Street
23   Brushes Estate
24   Hallyburton Road
25   Abbeyfield
26   Wolseley Road
27   Albert Road
28   Winn Gardens
29   Machon Bank
30   Firvale

with outer city locations, are shedding labour.
Mismatch between the demand for and supply of labour
is known to occur especially in the older industrial
estates and housing estates found in Scotland, Wales
and the North of England. Kirby, built to absorb
Liverpool's overspill, experienced an unemployment
rate of 4C per cent in mid 1982. In the outer
Glasgow area, unemployment rose dramatically between
1977 and 1978, reaching levels of 25 per cent in the
New Towns as well as in the city and old industrial
towns - only the suburbs performed rather better."

McGregor's (1979) analysis of urban unemployment
demonstrate this point. The focus of McGregor's
study was the Ferguslie Park estate, a public housing estate of about 1C,CCC people in the town of
Paisley which has traditionally enjoyed relatively
low unemployment rates. However, in Fergulie Park
the 1981 unemployment rate was 22%, a rate exceeding
that of many inner city areas. In Liverpool,
Webber's analysis of deprivation for the DoE (DoE,
1977a) confirmed the presence of deprivation in the
central part of that city, but also indicated the
emergence of deprived populations in housing estates
on the fringe of the city, particularly at Speke.
Thomas and Winyard (1979) examined income data from
the 1977 New Earnings Survey for individual counties, and concluded that low income families lived
both in urban and rural areas. The very poorest
lived in the counties of Cornwall, Devon, Powys and
Gwynedd. Shaw (1979) concludes that the dominance
of the urban poor in discussions of poverty arises
from their visibility and the threat to public order
which they are perceived to represent. In the light
of 1981, and the Handsworth riots of 1985 this
conclusion has some validity.

Although the plight of the rural poor may not
have entered the public consciousness, it has not
gone unnoticed by the academic community. Indeed,
Moseley (198C) prepared a paper on rural poverty for
the Social Science Research Council's Inner cities
Working Party which emphasised that the scale of
poverty in rural areas was underestimated, in part
because the nature of rural deprivation was mis-
understood and because the deprived in rural areas
are, by definition, more scattered. In an admit-
tedly crude analysis, Moseley showed that in 1971
there were more people living in administrative
rural districts (1C.57 million) than in the five
provincial English conurbations (8.4C million).
After examining a number of deprivation indices
Moseley (198C, 23) concluded, "compared with the

five Conurbations, rural England and Wales has more
elderly people and more households living in a
vulnerable sector of the housing market - privately
rented accommodation. In absolute and relative
terms it is (or was in 1971) only marginally better
placed in terms of unemployment." Part of the
difficulty in capturing the scale of rural poverty
is the deficiency of good measures since the census
includes few questions which capture the nature of
rural deprivation. There are, however, some common
features of urban and rural deprivation, neatly
summarised by Moseley (Fig. 4.6).

All of this evidence suggests that the commonly
held view of the distribution of deprivation is too
simple; whilst it is an urban phenomenon, and
whilst its symptoms are obvious in the inner city,
deprivation is also a suburban and a rural phenomenon. A somewhat different approach to the issue
has been followed by Donnison and Sotto (1980).
Instead of statistically carving up parts of individual cities they set out to identify the "Good
City" and the "Bad City", places which provide the
components of a "good" life for their residents,
particularly manual, industrial workers and the
elderly, and places which do not. Their conclusion
is that the good city is found in 14 towns of
"Middle England". Hall (1981, 60) summarises such
places thus: "by middle-class standards they are
probably not very desirable places to live in; they
actually contain more than their fair share of the
vulnerable groups, and they have fewer people with
'A' levels or equivalent. Yet they seem to provide
rather well for their inhabitants, their economies
demanding middle-range skills for which middle-range
incomes are paid ... matched with middle priced
housing, and with educational systems equipped ...
to give their children middle-level skills." And
the "Bad City"? Though the "Bad City", like the
"Good City" has its concentrations of manual, semi-
and un-skilled workers it is bad because its industrial basis is in decline, which is reflected in high
male unemployment. The conurbations and the towns
of Central Scotland are the archetypes.

These ideas on places as being good or bad ones
in which to live highlight, albeit in an intuitive
way, the centrality of well-being. They suggest too
that well-being is a perception as well as an ojbec-
tive state (see Pacione, 1982). As we indicated in
Chapter One, 'health' summarises for many people a
good quality of life. While we do not intend to
pursue these subjective dimensions of well-being or

Figure 4.6 Overlapping sets of problems in urban and rural areas

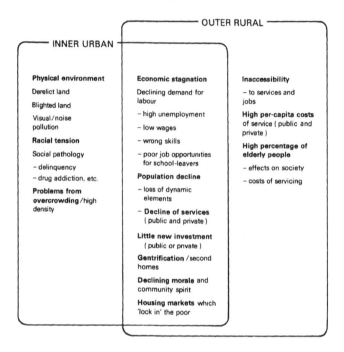

INNER URBAN	OUTER RURAL	
Physical environment	**Economic stagnation**	**Inaccessibility**
Derelict land	Declining demand for labour	– to services and jobs
Blighted land	– high unemployment	**High per-capita costs of service (public and private)**
Visual/noise pollution	– low wages	
Racial tension	– wrong skills	**High percentage of elderly people**
Social pathology	– poor job opportunities for school-leavers	– effects on society
– delinquency	**Population decline**	– costs of servicing
– drug addiction, etc.	– loss of dynamic elements	
Problems from overcrowding/high density	– **Decline of services** (public and private)	
	Little new investment (public or pnvate)	
	Gentrification /second homes	
	Declining morale and community spirit	
	Housing markets which 'lock in' the poor	

Source: M. J. Moseley (1980) Rural development and its relevance to the inner city debate, *Social Science Research Council, Inner City in Context, Paper 9,* figure 5.

health (see Eyles & Donovan, 1986), they significantly shape needs and demands for health care and
welfare in general. They should perhaps be more
fully treated for a truly integrated approach to
care policies. But the policies that we wish to
assess tend to ignore such perceptions. Indeed, the
policies underemphasise general well-being (objec-
tively defined) as opposed to a compartmentalised
view of need (for example, health care, transfer
payments, sheltered accommodation). This chapter
has related the who and where to the what. We now
wish to go on and link these with how. In other
words, how do those who need care obtain it? Part
of the answer to that question revolves round the
type of care available as the next chapter will
show.

Chapter Five

RESOURCE ALLOCATION POLICIES: INTEGRATING
PRINCIPLES AND SUBSTANTIVE OUTCOMES

We have tried to establish the case for adopting an
integrated approach to health as well-being. This
integration may take one of several forms. In the
context of welfare geography, it means examining the
basic question: (who decides) who gets what where
and how in its entirety. Indeed we have suggested
that the book in itself may be regarded as an
attempt to tease out some of the many relationships
and assumptions of that question. We reserve explicit consideration of welfare geography until Chapter
Six. In this chapter, we are primarily concerned
with the practice or policy implications of an
integrated approach. We do not pretend that such
integration has not been advocated or practised. In
fact, in determining the nature and quantity of care
(the what) to be allocated (how), policy has often
been based on an integrating set of theoretical-
conceptual premises which we have already encountered in Chapter Two. We intend to illustrate some
of these in this chapter, through a portrayal of
elements of the geography of the national health.
     Advocacy and practice have taken several forms.
Just as it was possible to identify three broad
categories of allocational mechanism so too we may
isolate related systems of integration or integrating principles. Integrated practice is always for a
specific purpose and from a specific perspective.
Policy is of course not a neutral technique for
enhancing improvements in health states or the
quality of life. Policy is shaped by those whose
practice it is and the social order in which that
practice occurs. All policy-making is social engineering of a kind with the engineering depending on
the values, aims and power of the 'engineers'. Not
all social engineering need be imposed from 'above'.

It is possible for a society to agree on its courses
of action which must then be implemented by the
establishment of an administrative apparatus to
alter conditions to achieve the desired state(s).
We suggest that this harmony of interests and integration of institutions are found in, established and
effected by socialist allocational mechanisms and
social orders.

We wish, however, to address three other types
of integration. First, there is emphasis on seeing
the problems of ill-health and ill-being in an
integrated way. We may illustrate this with reference to the New Poor Law of 1834 which saw poverty
and social ills as being the fault of individual
sufferers. The necessary policy was to limit the
claims of such individuals on the resources of
society. Thus the problem was morally defined while
the solution was ideally economically restrictive.
We may regard such integration as moral-economic
with close parallels to determinations established
by market allocational mechanisms. Secondly, the
solutions may be viewed in an integrated way, as
with the attempt to establish a framework for the
provision of social insurance and welfare services
begun in the early twentieth century, formalised by
the Coalition wartime government and instigated in
the 'consensus' politics of the post-1945 period.
Problems were seen as simply 'there' and state
social legislation was regarded as the prime method
of ensuring (engineering) freedom from ill-being.
This integration may be called social engineering
deriving mainly from bureaucratic interventionist
strategies. Thirdly, and in many ways as a reaction
to these strategies, we may identify financial
integration. Problems and solutions are still seen
to some extent as social and economic phenomena but
the main criterion is that of cost. The main principle is, therefore, cost containment. Integration
is at the level of budget, resulting is some unintended and disintegrating tendencies as the effects
of cost containment in one arena spill over into
another. There are parallels again with advocacy of
the market mechanisms and also at first sight with
the New Poor Law. Certainly both integrations are
meant to be economically restrictive but the financial integrationists usually lack the moral commitment to service (albeit narrowly conceived) of the
Victorians and Edwardians.

The New Poor Law and Public Health: Moral-Economic
Integration

The new Poor Law of 1834 was introduced because of a
crisis in resource allocation. By the 1830s the
Poor Law system dating from Elizabethan times was in
a muddle because each of the 15000 parishes were
independent and gave different kinds of assistance
to the poor. The parish was a deeply significant
entity and attempts to widen the basis of administration failed. It was to their parish, therefore,
that the poor looked for relief.

> The parishioner who became unable to main-
> tain himself or his family could expect to
> receive outdoor relief - a small pension,
> exemption from rates and assistance with
> his rent - or indoor relief, in the work-
> house or poorhouse. He could expect to be
> helped in sickness, infirmity and old age,
> in unemployment or inadequate wages. He
> could expect to be buried, his children to
> be apprenticed, his wife to be assisted in
> childbirth and his daughter helped to
> recover payments from the putative father
> of her illegitimate child. But he was
> secure of all this only in the parish where
> he had established a 'settlement'.
> (Henriques, 1979, 11)

The local assistance created few problems when the
population was settled mainly in sparsely peopled
rural parishes. But the British population was
growing rapidly from around 6 million in the mid-
eighteenth century to 14 million by 1834. These
demographic changes were exacerbated by economic
changes - industrialisation, enclosure, the decline
of cottage industries and in the early nineteenth
century a slump in wheat prices and a decline in
farm employment (see Digby, 1982). Particularly the
bad harvests coinciding with the war with France saw
the total cost of the poor rates to the country rise
from £1.5m in 1776 to £4.25m in 1802. It continued
to rise to £6.6m in 1813 and £9.3m in 1817-18.
After levelling off in the 1820s, it rose again to
£6.6m in 1831-2 and 1832-33 (see Bruce, 1971;
Digby, 1982). The total represented only a small
part of national income (probably no more than two
per cent), and it was associated with a rising
population so that per capita relief expenditures
actually fell from 12 to 13 shillings (1816 to 1819)
to 9 to 10 shillings (early 1830s). But such expenditure amounted to one-fifth of total national

expenditure and represented a rise of 62 per cent in
the poor rates from 1602-3 to 1832-3. Like the
rates of today, this expenditure loomed exceptionally large. At the time, it seemed to threaten the
economic foundations of society.

The Royal Commission on the Poor Law of 1832
had, therefore, several preconceptions (based
largely on Bentham and Malthus) on which to set
their findings. Their empirical investigations
enabled the drawing of pre-determined conclusions
and the production of a 'wildly unhistorical'
(Tawney, 1936, 269) and 'wildly unstatistical'
(Blaug, 1964) report. We must recognise the significance of the commercial age in which it was produced. It mattered not to the Commissioners that
the old Poor Law was a response rather than a cause
of population growth, under-employment and low
wages. Indeed, they avoided causation. Theirs was
a commercial age which saw property as the foundation of the social order and the labourer as something less than a full citizen. Such an age
replaced his parochial rights with a cash relationship and in the ideas of Bentham and Malthus found
distress to be caused by individual improvidence and
vice. Relief should therefore be administered not
merely to relieve but also to deter. It enabled the
overlooking of economically rather than morally
caused unemployment and of sickness as a cause of
poverty. It allowed the integrated treatment of
sufferers as moral degenerates as what Thompson
(1971) called the moral economy became replaced by
political economy. Economic criteria dominated
relationships but they were underpinned by strong
moral views.

Just as it is not our intention to provide a
complete historical overview of health and welfare
services, we do not intend to describe in detail the
operation of and amendments to the Poor Law. We
should note, however, that its two principles were
the workhouse test (ie the abolition of all out-
relief to the 'able-bodied' and their families whose
only relief would be to enter the workhouse in which
there was to be strict age and sex segregation) and
the principle of less eligibility (ie the condition
of inmates was to be made more miserable than those
of the lowest paid labourer). The importance of the
workhouse in its material deprivation and psy.cholo-
gical harshness must be recognised. It was an
influence that affected the attitudes of generations
of people to institutionalisation of all kinds,
including hospitalisation. These two principles

were meant to provide the new Poor Law with a framework within which all future decisions about poverty
and welfare could be made. But it did not work out
like that. Outdoor relief remained important,
particularly in industrial areas where there were
anti-Poor Law movements, connected with factory
reform. Indeed, in some areas workhouses were not
constructed until the 1860s (eg Lancashire, West
Riding of Yorkshire). And it is in the geography of
the Poor Law that we may find the issue of resource
allocation confronted.

In the new Poor Law, there is an apparent,
powerful central intervention in affairs of localities. It was not of course in the form of resources
that this intervention came but as direction and
advice. As Roberts (1960) notes, the central body
(Commission and then Board) came to possess a vast
accumulation of powers (legislative, judicial and
administrative). But local government was actually
strengthened because the central directive ensured
the formation of new local unions. As Digby (1982)
comments, local autonomy was gradually submerged
under expert advice and bureaucratic red tape which
led to the establishment of the 'poor law mind'
locally and nationally. Resources were raised
locally.

Indeed a major reason for the introduction of
the new Poor Law was to limit poor relief and rates.
The desire for economy in rating could be seen as
being complementary to depauperisation and the
restoration of self-reliance and independence;
except where the construction of special institutions like the workhouse were required. This
parsimony was another reason for delay in implement-
ing the Poor Law legislation in some localities.
While some delayed, other areas were quick to seize
the opportunity to reduce rates, particulary in the
southern counties of England. And East Yorkshire
reduced its poor relief expenditure by 13 per cent
in 1835 and by a total of 27 per cent from 1834 to
1837 (Henriques, 1979). But the new Poor Law led to
an amalgamation ('unionisation') of parishes. By
the end of 1835, 2066 parishes had been combined
into 112 Unions, covering about one-tenth of the
total population and one-sixth of the local rates of
England and Wales, and mainly in central and southern England. By 1836, there were 365 unions incorporating 7915 parishes. Progress slowed thereafter
as the Commissions tried to unionise the industrial
north and came up against other local unions. But
by the end of 1839 only 799 parishes were outside

Poor Law unions, there being some 583 unions, incorporating 13691 parishes. By 1871 the process was
about complete with the 15000 parishes in some 600
unions (see Henriques, 1979). They endured until
1930 when their functions were absorbed by the local
authorities under the Local Authority Act of 1929.
The Board itself was replaced in 1871 by the Local
Government Board which became The Ministry of Health
in 1919 (Bruce, 1971).

Unionisation was not a politically neutral act.
Local property owners helped in the creation of the
poor-law unions. Brundage (1972) argues that the
great landowners had the unions gerrymandered to
suit their own interests. They drew the union
boundaries around their estates as in Northampton-
shire with the unions at Potterspury (Duke of
Grafton), Aynho (Cartwright) and Daventry
(Knightly). The largest town was left consisting of
four town parishes and thirteen rural parishes
without great landowners. Rural ex officio guardians were used to overwhelm the elected representatives of the parishes of Northampton. Dunkley
(1973), however, disagrees showing that yeomen
farmers or town shopkeepers and artisans could
secure control of the Boards of Guardians. Else-
where, past practice and geography influenced
unionisation. An existing Local Act Union at
Shrewsbury was entirely surrounded by the new Poor
Law Union Atcham, while on Anglesey, the five
eastern parishes were attached to Caernarvon and 15
to Bangor because of existing links across the Menai
Straits (Henriques, 1979). Further, the local
economy could affect the introduction of the system
with the mixed economy of the North East being more
stable than the textiles of Lancashire and the West
Riding, where it was futile to see the workhouse as
a way of providing welfare during trade slumps (see
Digby, 1982).

Control of Board of Guardians was of course
vital because on this depended the level and nature
of resource allocation. The intention of wresting
the Poor Law from large landowners was only partly
successful. In rural areas, control was often as
not obtained by large tenant farmers rather than
artisans, shopkeepers and smallholders, and they
were intent rather on spreading the burden of their
wage bill as the existence of relief meant that they
could pay low wages, and pay only when work was
carried out. In urban areas, elected representatives could gain control. But control determined
where the rate should fall. All were keen to reduce

the amount paid as rates. Thus as the Board had no power to compel the levying of rates for a workhouse its more constructive projects (schools, medical services) were precluded. Local taxation gave the local Boards of Guardians an independent power base to determine the level of resources and manner of allocation. The parish remained the basis for rating and settlement until the 1860s. Such a system is perhaps acceptable in a static society with a reasonable areal balance between propertied and poor. This had however ceased to be the case in the nineteenth century. And while the new law succeeded in reducing national expenditure on Poor Law relief from an average of £6.3m pa between 1813 and 1834 to £5.2m pa between 1835 and 1860 (Mitchell and Dean, 1962), the burden was unequally shared. The areas with most paupers were often those with impoverished rate payers within Newcastle, for example, a considerable proportion of the rates being drawn from those little removed from pauperism themselves (McCord, 1978). It also proved difficult to tax the new commercial and industrial wealth so that householders shouldered heavy rates. In general we may note, therefore, a failure to reform the basis of financing welfare. It was feared that a wider geographical basis would increase pauperism and the rates. But the failure of the local basis of funding to cope with the demands emanating from economic and trade changes as for example in Lancashire and London led to the establishment of common funds. By the 1860s, contributions to the common fund were assessed on ability to pay as estimated by the value of occupied property rather than past levels of pauperism (see Digby, 1982). And by that time too specific expenditures, including the care of the sick, were charged to the common fund.

As F B Smith (1979, 351) has noted "consideration of the old and infirm entered the planning of the New Poor Law only as an inescapable afterthought." Medical services probably deteriorated after the 1834 Act, especially in the period up to 1842 (see Hodgkinson, 1967; Flinn, 1976). Medical attention was seen as being possible to provide through loans which the patient would repay when s/he recovered. While probably this was unenforceable, the parsimony meant that the patient lost his/her right to choose their own medical attendant. Outdoor pauper sick were treated at the discretion of guardians who had so many pauper tickets at their disposal. Once the tickets were used, treatment ceased. Doctors appointed to the unions were on part-time contracts and were seen as the salaried servants of the Board rather than independent agents. The contracts were parsimonious. The contract for the scattered parishes around Tunbridge Wells was reduced by the Commissioner from £470 per year to £250 while a tender to treat 7,902 paupers in Balsham and Linton in Cambridgeshire was won with a bid of £90. Balsham with 3614 paupers had pre-viously been served singly for £105 a year.

Complaints of neglect and ill-treatment and the scale of the problem - Gregg (1982) suggests that 30 per cent of the pauper population was sick - led to the Commissioners pressing boards to appoint quali-fied doctors at reasonable wages based on the popu-lation served. But many areas resisted. For example, the smaller farmers of South Wales resented paying their officials more than they earned them-selves and providing medical attention they them-selves could not afford (see Henriques, 1979). But from 1642 unions were divided up into medical dist-ricts which were not to exceed 15000 acres or 15000 population. These limitations on size were often ignored so that the cost of providing the service could be effectively reduced. The Poor Law medical service remained haunted by the principle of less eligibility and the parsimony of the ratepayers. Respectable people who applied for medical help were stigmatised as paupers and deprived of civil rights. All applicants had to obtain an order to see the doctor from the board's official and just as these officials saw their monopoly of relief being eroded by doctors' prescriptions, doctors also saw their fledgling profession being shaped by others. The money for the service continued to be massively inadequate with Flinn (1976) estimating that in 1640 only £150,000 out of a total Poor Law expenditure of £4.5m was spent on medical relief. In 1871 less than £300,000 was spent out of a total of just under £6m. Indoor treatment in workhouse infirmaries grew rapidly but they were places of overcrowding and squalor, staffed by untrained nurses. Indeed it was not until 1897 that pauper nursing was prohibited. The squalor of the hospitals has been documented by McKeown and Brown (1955-6), Woodward (1974) and F B Smith (1979), although there is disagreement over whether Poor Law and municipal infirmaries did their patients more harm than good.

While services remained economical, improvements and reforms were carried out. In part, these occur-red under the aegis of the public health movement

and Public Health Act of 1848. A direct link between poor medical care and these other events is
found in the person of Chadwick, secretary of the
Board and author of the report on the living   conditions of the working classes. This link may be
one of the reasons why public health measures, ie
sanitary reforms, were not more fully supported by
the working class. "Those who held Chadwick responsible for the new Poor Law would hardly credit him
with benevolent motives in public health, and his
persistent refusal to acknowledge that disease might
be caused by poverty could only confirm their suspi-
cions" (henriques, 1979, 147). Indeed, the impetus
of sanitary reform came from the equation of dirt
with disease, with dirt being seen as a moral as
well as medical matter. But it was the effects of
the typhus epidemic on Poor Law populations in
1637-8, the reports of medical officers on conditions in such cities as Liverpool and Leeds,
Chadwick's own report on sanitary conditions and the
1854-5 Health of Towns commission that culminated in
the Public health Act of 1848. This was passed
against much opposition of the anti-centrist kind.
It, however, enabled rather than compelled. London
had excluded itself from the Act and by late 1853
only 164 places had adopted it. Many towns, like
cholera-ridden Newcastle, instituted ineffective
local acts to forestall imposition of any compulsory
clauses or moves to adopt the central Act (F B
Smith, 1979). Local political complexion influenced
whether the act would be adopted with reform in
Leicester and Leeds stopping with Tory rule and
starting when Liberal Dissenters took control
(henriques, 1979). Hancock (1973) shows how the
increase in shopkeepers and small business interests
in the municipal affairs of Leeds and Birmingham led
to parsimony and refusal to tackle the sanitary
problem. In fact, local interests - those who
funded and allocated resources to sanitary reform -
succeeded in suppressing the General Board of Health
in 1854, with The Times commenting

> If there is such a thing as a political
> certainty among us, it is that nothing
> autocratic can exist in this country. The
> British nature abhors abolute power... The
> Board of health has fallen... We all of us
> claim the privilege of changing our doc-
> tors, throwing away their medicine when we
> are sick of it, or doing without them
> altogether whenever we feel tolerably
> well... Esculapius and Chiron, in the form

of Mr Chadwick and Dr Southwood Smith, have
been deposed, and we prefer to take our
chance of cholera and the rest than be
bullied into health (quoted in Marshall,
1970, 21).

And take their chance they could as epidemic followed epidemic culminating in the cholera one of
1866 which killed some 20,000. These outbreaks, the
fact that local efforts appeared to do little to
ameliorate conditions and the economic costs of a
debilitated workforce led to greater agitation for
change. In 1866 a sanitary act was passed that
required local authorities to take action but local
boards of health were not established throughout the
country until the Public Health Acts of 1872 and
1875, the latter codifying and systematising existing legislation. As John Simon, the reformer perhaps most responsible for the national extension of
sanitary reform, said "the fact that Preventive
Medicine has now been fully adopted into the service
of the State is indeed the end of the argument"
(quoted in Bruce, 1971, 134). In a way it was
because of sanitary reform and increasing medical
comprehension of disease causation that mortality
rates fell.

   But the central significance of the Poor Law and
poor-law mind remained demonstrated by Simon's own
resignation from the local government board which he
saw dominated by poor law practice. Other reforms
were occurring within that practice. Sanitation was
seen as not being sufficient improvement. Infection
and contagion called for the isolation and improved
care of patients. In 1865 a Medical Officer of
Health was appointed to the Poor Law Board and
unions were encouraged to form 'sick asylum districts', large enough to support hospitals to which
the sick could be removed from the workhouses to
form what Bruce calls an embryonic national hospital
service. The greatest improvement came in London
where the Metropolitan Poor Act of 1867 established
a metropolitan common poor fund which pooled the
resources of the London union and parishes and
established a Metropolitan Asylums Board. The
Board built and maintained hospitals for infectious
cases, infirmaries for the non-infectious, asylums
for the mentally ill and dispensaries for those not
requiring in-patient treatment. With its teaching
hospitals, London's poor were well-served with
medical treatment and the London infirmaries were
gradually taken away from poor law control. The
importance of the voluntary sector and this change

of control are shown in Table 5.1, which also shows
how well-served London was compared with the rest of
England and Wales.

The poor law hospital remained an important part
of the service until the changes instigated by the
Local Government Act of 1929. Although municipal
hospitals were sanctioned and encouraged from 1875
onwards, they were slow in being developed. Poor
law hospitals were, therefore, thrown open to those
that required treatment. Admission was legalised
without any question of poor relief in 1883 and two
years later such admission no longer disenfranchised
people. These changes saw the ideas of the new Poor
Law finally overturned in the treatment of the sick
and poor.

5.1 Number of Beds for the Physically Ill per 1000 Population 1891-1938

hospital Type	1891	1911	1921	1938
Voluntary				
London	1.71	1.73	1.85	2.38
Provinces	0.85	1.06	1.40	2.04
Poor Law				
London	0.76	2.64	2.53	-
Provinces	0.34	0.75	0.58	0.24
Municipal				
London	-	-	0.03	0.24
Provinces	-	-	0.04	0.74

Source: Halsey 1972, 350

It was recognised that by 1905 they had become
practically state hospitals. Their significance is
illustrated in Table 5.2 which shows that poor law
infirmaries and sick wards provided 64.5 per cent of
hospital beds in 1891, 52.7 per cent in 1921 and
even 19.9 per cent in 1938. This more liberal and
less penal attitude to the sick poor pre-dates the
reforms to welfare provision heralded by the 1905-9
Royal Commission and initiated by the Liberal
government in the years before the first world war.

It is not our intention to show how and why the
Poor Law legislation was finally killed off. Nor do
we intend to show in historical sequence what happened next. We may conclude with the Webbs who
pointed out that the principles of 1834 "had been,
almost unawares, gradually abandoned in practice"
and that "there had grown up, during the preceding
half century, an array of competing public services
which were aiming, not at the prevention of pauper-

5.2 **Number of Beds for the Physically Ill and their Percentage Distribution in Voluntary, Poor Law and Municipal Hospitals, 1891-1936**

Hospital Type	1891		1911		1921		1936	
	No	%	No	%	No	%	No	%
Voluntary	29520	26.2	43221	21.9	56550	24.7	87235	33.2
Poor Law – infirmary[1]	12133	10.6	40901	20.7	36547	16.1	7909	3.0
sick ward[2]	60778	53.9	80260	40.6	83731	36.6	44556	16.9
Municipal	10319	9.2	33112	16.8	51728	22.6	123403	46.9
Total	112750	100.0	197494	100.0	228556	100.0	263103	100.00

[1] separate institutions
[2] in work houses

Source: Halsey 1972, 349

ism, but at the prevention of the various types of
destitution out of which pauperism arose" (quoted in
Bruce, 1971, 115).   But less eligibility had created
a climate of opinion - the poor law mind - averse to
new ways of dealing with poverty and ill-health.
But the principle failed because the poor remained a
constant element in a rapidly growing population.
It failed too because its system of resource allocation was based on variable local taxation with the
the additions by the Acts of the latter part of
nineteenth century giving a picture of confused
responsibility and uneven burden.

> Poor Law charges in 1926 varied in the
> county boroughs from 5d in the pound at
> Blackpool to 16s 5d in the pound at
> Gateshead; in rural unions from 2 and one
> half d in the pound at Howden to 5s 4 and
> one half d in the pound at Pontardawe;
> within the same county they varied so that
> in Brecon 11 and one farthing in the pound
> was the rate at Brecon Union and 7s and
> half a penny at Crickhowell (Gregg, 1982,
> 499).

It was replaced by a practice which gained confirma-
tion from the studies of poverty and ill-health of
Booth and Rowntree and which was enhanced by pragmatic politicians and administrators of the reluctant collectivist type (see Chapter 2). Poverty and
welfare needs became seen in terms of individual
characteristics like low income, ill-health, old
age, and large families. Further, an integrated
approach to resource allocation based on the view
that welfare recipients were morally degenerate
could not survive the creation of a more democratic
state brought about by agitations for and eventually
extensions of the franchise. While practice tended
to deflect and dilute the principles and procedures
of the moral-economic integration of the Poor Law,
it was not until 1948 that the National Assistance
Act stated "the existing poor law shall cease to
have effect."

The Welfare State and Planning: Socially Engineered Integration

In moving to the period after the Second World War,
we are not implying that parts of the development of
the welfare and health services can be omitted from
consideration. It is rather that particular elements illustrate well the combination of principles
and practice that we have defined as integration, ie

an approach to welfare that is holistic in at least
one regard. While in our first example, it was the
problems of poverty and ill-health that could be
viewed in an integrated way through their cause, in
this second case, the integrating element is seen as
the solution in the provision of universal benefits
to the population. This is not to say that a moral
dimension is absent. But in this case, the appeal
is to a far more sophisticated argument than that
present under the New Poor Law or our third example,
cost containment. In this instance, it is an appeal
based on the notion of citizenship. All employed
members of the society contribute with benefits
accruing without means testing to those that qualify
for assistance through unemployment, sickness,
disability, low wages and so on. As we shall see,
qualification is determined by bureaucratic
criteria. There is also present an intergenera-
tional integration which may be seen particularly
clearly in the provision of pensioners. The present
generation of wage-earners provide the contributions
to meet the pensions of past workers and will rely
on their children to meet their own pension demands.
To anticipate, these measures for integration contain within themselves the seeds which were to
destroy the consensus over solutions, namely the
'interference' in individual lives apparently
demanded by bureaucratic provision, collective means
'interferring' with individual freedom of choice to
spend (or save) as individuals saw fit, and through
an ageing population, the perceived escalating cost
of health and welfare services.

Further, in isolating these examples, we do not
wish to suggest a segmented view of welfare development, which must be seen in terms of the continuous
interaction and interpenetration of ideas and practices. Thus, for example, the voluntary practice of
the Friendly Societies and the regulations issued in
Bismarck's Germany concerning different types of
employment greatly influenced the introduction of
compulsory social insurance by Lloyd George's
Liberal government (see Marshall, 1970). But the
inter-war period, so formative for the post-1945
welfare state, was perhaps stronger on practices
rather than ideas. Even before this period, it was
the great social research documents of Booth and
Rowntree on the extent and conditions of poverty
that highlighted the impact of social circumstances
rather than personal attributes on hardship and
ill-health. Indeed the laissez-faire theories of
the 1920s and 1930s appeared to exacerbate social

ills. The ideas of Keynes on demand management and
employment were not fully articulated until 1936 and
in fact became part of the war-time efforts to
establish a framework for reconstruction.

We have already noted in our discussion of
justice and allocation that Keynes and Beveridge
were reluctant collectivists. But that reluctance
was overcome by a pragmatism that pervaded the
initial construction of the welfare state - a term
first used in the 1930s to distinguish between the
policies of the democracies and the 'war states' of
the dictators (see Stevenson, 1984). In this decade
'planning' became a notion not just reserved for
townscape but now included the management of social
and economic affairs. This 'middle opinion' was
crystallised by such pressure groups as Political
and Economic Planning and the Next Five Years Group
which demanded co-ordinated policies based on substantive research (see Marwick, 1966; Pinder,
1981). In 1933, harold MacMillan wrote:

Planning is forced upon us ... not for
idealistic reasons but because the old
mechanism which served us when markets were
expanding naturally and spontaneously is no
longer adequate when the tendency is in the
opposition direction (quoted in Stevenson,
1984, 328).

As with the Great War, the Second World War with its
necessary extension of state control, confirmed the
role of the state in a spirit of communal solidarity. As Milward (1977) points out, it was at the
height of the war that the education and town planning systems were being reformed and the welfare
state widened and made more generous. In many ways,
these reforms and reconstructions typify the integrated approach - planning, conceived as state
intervention, direction or provision, was seen as
the solution. This planning was by definition a
pragmatic response but this must not imply that it
was not ordered. With 'planning' went the idea of a
more rational, ordered treatment of social questions
- a technocratic, or social engineering approach.

The social engineered, integrated approach may
be seen in the wartime and immediately post-1945
legislation that was so influential in the creation
of the planned and welfare state. While it is
beyond our remit to review this legislation, we must
note some of the town planning as well as welfare
reports of the period. Hall (1974) has suggested
that the 1940 Barlow Report had an impact on British
urban and regional planning that can never be over-

estimated. The Report emphasised the relationship
between regional decline, migration and metropolitan
growth, regarding the problems of depressed areas
and of excessive growth as different aspects of the
issue of population distribution. It tried to
indicate the associated problems of housing and
health problem areas and, in the influential minority report by Abercrombie, recommended controls on
industrial location. While it does not concern us
that the policy impact of Barlow was slight (see
Cullingworth, 1973) and that changes came because of
the need for physical planning controls, the Barlow
Report was in part responsible for other documents
including Abercrombie on the planning of London and
Keith on new towns, both of which resulted in significant improvements in housing and living conditions
and the general well-being of a significant proportion of the population.

The Report of greatest interest to us, however,
must be that of Beveridge, who described the Plan
thus:

> The Plan for social security is put forward
> as part of a general programme of social
> policy. It is one part only of an attack
> upon five giant evils: upon the physical
> Want with which it is directly concerned,
> upon Disease which often causes that Want
> and brings many other troubles in its
> train, upon Ignorance which no democracy
> can afford among its citizens, upon the
> Squalor which arises mainly through hap-
> hazard distribution of industry and popula-
> tion, and upon the Idleness which destroys
> wealth and corrupts men, whether they are
> well fed or not, when they are idle (Gt.
> Britain, 1942).

Social security formed but one part of the integrated approach which also included then policies
for full employment, family allowances and national
health service (see Doyal, 1979). Beveridge's
approach was to blend the experience of the past
(including the commitment of working people to
national survival which should result in government
provision of services) with a more radical approach
towards existing vested interests; comprehensive
social planning and cooperation (the basis of moral
integration) between voluntary and public sectors
and the individual and the state. On this last
point, the Report argued that people were not being
given something for nothing or being freed from
their personal responsibilities. Individuals were

to be given the freedom and encouragement to "win
for themselves" something but the national minimum
(Great Britain, 1942). But the assurance of this
national minimum was the bedrock of the Plan. And
despite opposition from both conservatives like
Churchill ("It is because I do not wish to deceive
the people by false hopes and airy visions of Utopia
and Eldorado that I have refrained so far from
making promises about the future") and socialists
like Bevin (the Plan is "the culmination of ideas on
social services over the last 40 years ... a co-
ordination of the whole of the nation's ambulance
services on a more scientific and proper footing")
the assumptions concerning family allowances, the
health services and unemployment were accepted in
principle (quoted in Bruce, 1971, 305, 309).

In all this, however, Beveridge saw the state
playing a necessary but minor role. The state
should only do those things which it can do alone or
better than any local authority or private citizen
singly or in association (see George & Wilding,
1976). Beveridge, like Keynes and so much 'middle
opinion', was in favour of social reform so long as
the existing structure of society remained fundamentally the same. "In fact, he was for social reform
precisely because it would allow the existing order
to continue essentially unchanged" (Kincaid, 1973,
48). And this point has remained accurate up to and
including the efforts of the last Labour government
of 1974 to 1979. State involvement has greatly
increased but the capitalist structure of the
British economy and polity has remained largely
unscathed (see below). Indeed it is not just the
establishmment of welfare within a reformed capitalism but the whole ethos of the post-1945 consensus that is summarised by Addison (1975):

> In general, the reform programme originated
> in the thought of an upper middle class of
> socially concerned professional people, of
> whom Beveridge and Keynes were the patron
> saints. To render capitalism more humane
> and efficient was the principle aim of the
> professional expert. In World War II the
> humane technocrat provided a patriotic
> compromise between Socialism and
> Conservatism which virtually satisfied the
> desire of the Labour Party for social
> amelioration, without in any way attacking
> the roots of exploitation and injustice.

Part of Beveridge's Plan was the establishment
of a national health service. Again our purpose is

not to replay the debate over its establishment (see
Navarro, 1978; Watkin, 1978; Pater, 1961). It is
intriguing to note that the debate over medical
services, carried out somewhat independently of the
wider welfare issues, echoes our own concerns for
health rather than sickness services. As the 1937
Report by Political, Economic Planning on the health
services commented.

> The really essential health services of the
> nation are the making available of ample
> safer fresh milk to all who need it, the
> cheapening of other dairy produce, fruit
> and vegetables, new accommodation to
> replace slums and relieve overcrowding,
> green belt schemes, playing fields, youth
> hostels and physical education, social
> insurances which relieve the burden of
> anxiety on the family and advances in
> employment policy which improve security of
> tenure or conditions of work and, finally,
> education in healthy living through train-
> ing and propaganda (quoted in Watkin, 1978,
> 4).

Such an improvement in health services may have been
possible if local (or regional) control had been
instituted because it may be argued that local needs
may be best deduced and met locally even if resources must be obtained nationally. Such institution
may well be impossible in such a centrist state as
Britain, despite the advocacy of the regionalisation
of health services on the bases of health centres
and hospitals as early as 1920 by the Dawson Report.
The recommendations of this report were, however,
significantly affected by the power and vested
interests of the medical profession which was to be
allowed to continue in private practice (see
Navarro, 1978). This power was used during the
discussions leading up to the National Health
Services Act of 1946 and the appointed day of 5 July
1946 when the service actually began. This power in
part prevented the establishment of a regional basis
for health care in which local authorities would
play an important coordinating role and substituted
a functional basis for the organisation of services.
This medical dominance - often billed as Bevan
versus the doctors - have been well-documented (see,
for example, Navarro , 1978; Eckstein, 1958;
Pater, 1961), although which side 'won' or obtained
more concessions remains a moot point. The presence
of medical power is, however, undeniable and may
have been sufficient to affect the attempts made to

establish an accountable health service despite the
great class pressure for change from which Bevan
drew great strength (see Corrigan, 1977; Doyal,
1979).

Indeed it is probably fair to say that the
conservative hold that this dominance exercise is a
major contributing factor to the unrevolutionary
character of the national health service. As the
White paper commented: "there is no question of
having to abandon bad services and to start afresh.
Reform in this field is not a matter of making good
what is bad, but of making better what is good
already" (Great Britain, 1944). Of Bevan, Willcocks
(1967, 104) comments, he "was less of an innovator
than often credited: he was at the end, albeit the
important and conclusive end of a series of earlier
plans. he 'created' the National Health Service but
his debts to what went before were enormous." On
medical care in general, Navarro (1976, 47) has said
that "general practitioners continued practising in
the same setting as before, and retained the much-
desired autonomy which, in the long run, has very
seriously hurt general practice in Britain ...
Within the medical profession those who have bene-
fited most have been the consultants, who made
substantial gains, both in income and power, with
the implementation of the NHS Act." And finally and
utilising a distinction first made by Turshen
(1977), Doyal (1979) argued that in 1946 Britain
nationalised rather than socialised its medical
care, with the state taking over responsibility for
organising such care in the same way as it took over
the mines or the railways. Few extra resources were
invested and power remained firmly in the hands of
those who had always been in control - the doctors
and the administrators. But these comments should
not mask the aims and achievements of the NHS which
set out to be "publicly sponsored service ... available to all who want to use it" (Great Britain,
1944). Indeed as Marwick (1982) notes, the NHS was
to be a monumental expression of the principle of
universality. Treatment was not dependent on contributions and the service was open to all and, for
most items, free at the point of service.

The functional division of services was a tripartite scheme of local authority services (home
visiting, child welfare), family practitioner services (FPS) including physicians, dentists and
opticians, and hospitals. FPS were excluded from
local authority control as this was deemed inappro-
priate where specialised professional judgements

would be involved. The NHS inherited an uneven
distribution of GPs and has attempted by defining
areas as designated, open or restricted to prevent
new doctors establishing themselves in over-doctored
areas and encouraging them to move to under-doctored
parts - bureaucratic intervention with a manifest
benign intention. But as Butler (1973) notes, this
policy has not been carried out forcefully or in a
coordinated manner, concluding by saying that the
distribution of GPs remained largely unaltered over
the last 30 years, with those areas having no difficulty in the past being still relatively well supp-
lied and those with serious shortages having a
fairly long history of manpower problems. There
have, however, been changes. Comparing average list
sizes between 1966 and 1973, we note that few places
that were the worst in their region at the earlier
data remain so at the later one. Intriguingly,
however, the best places, ie those with the lowest
list size are virtually all the same (eg Bournemouth
in the South East, Kesteven in the East Midlands,
Herefordshire in the West Midlands). Further as OHE
(1979) notes, the number of patients in 'under-
doctored' areas lessened by 39 per cent from 40m. in
1963 to 24m. in 1977. This represents a drop from
85 per cent of the total population of England and
Wales to 46 per cent in 1977. But just under 30
years of bureaucratic intervention still left nearly
half of the population in under-doctored areas.
Further, this socially engineered approach to FPS is
rather limited. There has been little or no intervention in general dental services, with Cook and
Walker (1967) and Bradley et al (1978) finding a
strong association between the distribution of
dentists and the social class of an area.
    We should further note that FPS has been the
'Cinerella' part of the NHS in general or at least
until the mid-1970s. In 1948, FPS accounted for one
third of the NHS budget. In 1974, this had declined
to fractionally over one fifth. Its slow rise since
this time to the present will be discussed below.
It is in fact the third element of the tripartite
scheme - the hospitals - that takes the bulk of NHS
funding and this perhaps mirrors the power of hospital consultants in the profession at large. The NHS
established, not as originally suggested regional
committees to run the hospitals, but ad hoc hospital
boards. Fourteen such boards were established in
England and Wales with each being centred on the
medical faculty of a university. Even then the 36
teaching hospitals were given a special autonomy

with their governors being appointed directly by the Minister. As Richard Crossman, Secretary of State for social Services (1968-70) commented on the regional hospital boards (RHBs)

> The RHB is the most perfect example of self-perpetuating oligarchy since the Persians' rule by Satraps ... Consultants are the most powerful autocrats in the world to order about chairmen of RHBs. A real iron law of oligarchy with the minister appointing the RHB and the consultants appointing themselves ... What chance is there of a shift of money to the community services or long-stay hospitals? (quoted in Navarro, 1978, 48).

Indeed, the consultants had won for themselves the right to use NHS facilities for private practice as well as great influence on decision-making and administrative bodies in the NHS. From its inception, the planning process within the NHS was dominated by a particularly powerful and partial elite (see Illsley, 1980; Eyles and Woods, 1983).

But consultant power was not the most visible problem of the new NHS, in hospitals or in general terms. This problem was financial as demand for services seemed to rise inexorably. The original estimate for 1948-9 of £176m. turned out to be £225m. Bevan argued that

> The rush for spectacles, as for dental treatment has exceeded all expectations ... Part of what has happened has been a natural first flush of the new scheme, with the feeling that everything is free now and it does not matter what is charged up to the Exchequer. But there is also, without doubt, a sheer increase due to people getting things they need but could not afford before, and this the scheme intended (quoted in Klein, 1983, 33-4).

Public expectation, itself depending on confidence in medicine, levels of education, rising standards of living and diminishing tolerance of pain and disability (see Brotherston, 1979) does indeed influence demand. But later evidence, particularly that presented in the Guillebaud Report, found that the rising cost of the NHS had been kept within narrow bounds (Great Britain, 1956; Abel-Smith and Titmuss, 1956). In fact, the proportion of national income spent on the NHS actually fell from 4.08 per cent of GNP in 1950 to 3.53 per cent in 1954 (OHE, 1984). Putting it somewhat differently, the NHS

accounted for 11.8 per cent of UK public expenditure
in 1950 (a proportion not seen again until 1977) and
8.8 per cent in 1953 (its lowest figure ever in the
post-1945 period) (OHE, 1984). This evidence was
not, however, considered at the time. Post-war
economic austerity and the costs of British involvement in the Korean war meant that charges for health
services were introduced in 1951. It was hoped that
such charges would also help curb demand in an era
which austerity and Cripps rather than reform and
Beveridge/Bevan dominate (see Morgan, 1984). In
fact, the financial innocence of the early NHS was
recognised by the Guillebaud Committee which on
balance favoured the retention of prescription
charges. And as Klein (1983, 35) notes "the logic
of the NHS's commitment to providing a free service
and so 'universalising the best', in Bevan's phrase,
ran counter to the logic of its dependence on the
Treasury and tax revenue for funds". Each new
demand for provision by the health and welfare
services would create an expenditure demand. The
proposed integrated solution - universal benefits -
ran up against the cost constraints of financial
provision. In such circumstances, with limited
funds and apparently infinite demand and need (and
how else should it be in a society increasing in
prosperity and wedded to a notion of justice based
at least on citizenship rights), eligibility became
an important matter. Fine tuning the welfare state
to meet new health and welfare requirements required
a mass of bureaucratic interventions and infighting
over funds - the very opposite, it would seem, to
'planning' on 'technocratic' criteria. With the
recognition of supply of a service helping to generate further demand came some of the seeds that would
lead to the partial destruction of the welfare state
in the name of choice and efficiency (see below).
    It is also interesting to note that the
Guillebaud Report also showed that the cost of the
hospital service was rising, mainly caused by the
growth in staff numbers, especially nurses and
auxiliaries, while capital investment in the NHS
actually declined in spite of the fact that 45 per
cent of all hospitals had originally been erected
before 1891. It is to the problems of hospital
provision that we now wish to turn as it illustrates
very well certain features of rational planning and
bureaucratic intervention. Such provision became
one of the most planned features of the welfare
services, although it became less and less integrated into the total pattern of care available (see

Chapter One) as the post-1945 period advanced.
Indeed such integration received a mighty setback
when in 1951 the Ministry of Health lost its responsibilities for housing. Hospital services demonstrate how 'planning' became emphasised rather than
integration, particularly in the form of the co-
ordination of services in specific sectors.

Just as medical care was discussed and planned
somewhat independently of welfare issues so too were
hospital services. During the war (in 1941) a
massive survey of hospitals was launched. This
showed an inheritance of some 0.5m. beds in 2,600
hospitals, many of which were small voluntary hospitals. Forty five per cent had been originally
constructed before 1891 and 21 per cent before 1861
(see Ministry of Health, 1945). It was found that
many were substandard and that the hospital building
programme had failed to keep pace with demographic,
social and medical change. So there already existed
in 1946 a complex if inadequate infrastructure (see
Abel-Smith, 1964; Pinker, 1966). But because of
variations between places, Grey and Topping (1945)
were surely correct when they concluded that any
exact measure of the adequacy of hospital services
was almost impossible. We may in fact note the
differences between North West England and London
which seem to predate the RAWP findings by some 30
years. In the North West,

it must be recognised that the existing
hospitals, considered as buildings, fall
far short of a satisfactory standard.
Indeed, considering the high place which
England takes in the medical world, perhaps
the most striking thing about them is how
bad they are. This is less surprising,
however, when one realises how old they
are. The number of new hospitals built
during this century is surprisingly small
and the number built in the inter-war
period very small indeed.

In London and the surrounding area, while quality
and quantity of hospital accommodation were seen as
deficient,

in many areas, however, wartime extensions
of the Emergency Hospital Scheme will
provide relief at least of a temporary
kind; and in London itself (apart from the
southern fringe of the county) it may be
increasingly found that there is too much
accommodation rather than too little. For
these reasons ... the question of hospital

building must be approached with circumspection.

The main planning tool as far as hospital services were concerned became the bed norm (standard
rate of bed provision per population). The Survey
regarded the key element in establishing bed norms
as consultant services, as they had clinical responsibility for secondary care of patients. Proposals
were made for consultant staffing and beds for each
speciality and such services were seen as being
provided in hypothetical average districts with
populations of between 100000 and 120000. Optimal
bed provision was established as 5 to 7 per 1000
population in England and Wales and 8 per 1000 in
Scotland. But existing infrastructure meant that
excluding mental facilities there already existed 7
beds per 1000 population.

The Survey provided the background to attempt to
establish the 'needs' of communities in terms of
bed norms. These attempts are themselves interventionist strategies in which rational planning and
criteria established external to actual or felt need
by experts are central (see Chapter Two). Studies
by Nuffield Provincial Hospitals Trust (NPhT) in
Scotland surveyed the use of hospital facilities by
residents (NPHT, 1946; 1950), assuming that utilisation accurately reflected need, perhaps a fair
assumption in the days before the recognition of the
clinical iceberg of illness (see Last, 1963). The
studies assumed a bed occupancy rate of 85 per cent
(a sound assumption given that in 1959 this rate was
as high as 87 per cent) and estimated that the
number of beds required was 5.0 per 1000 population.
Later studies of Norwich and Northampton (NPHT,
1955) and Reading (Barr, 1957) estimated the number
required to be even lower at around 2 per 1000.
Further studies in the less favoured North, however,
established required provision somewhat higher, eg a
critical range of 3.88 to 4.14 acute beds per 1000
in Teesside, compared with actual provision of 4.07
(Airth and Newell, 1962), 2.56 in Barrow-in-Furness
(Forsyth and Logan, 1960), while Logan (1964) argued
that there existed 'unmet need' in certain areas
because of restrictive admission policies, particularly for geriatric and maternity cases.

It is also possible to see these studies as a
background to the most comprehensive attempt to plan
hospital provision (Ministry of Health, 1962). The
Hospital Plan was devised in circumstances of optimism over national prosperity but a recognition that
resources available to health care would be finite.

Indeed, the recognition of the 'bottomless pit' of demand meant that greater emphasis needed to be placed on efficient provision. But capital expenditure which had been held firmly in check because of the housing and educational programmes as well as defence expenditure was to be increased. This expenditure would not only placate industrial and medical lobbies but would also be reformist as it could alter the spatial distribution of hospitals in favour of the more deprived regions. As Mohan (1984a, 282) remarks, "the announcement of the Hospital Plan saw the state simultaneously attempt to find a new means of restoring profitability for private capital, satisfy the insistent medical lobby for more capital expenditure, and improve social provision for the nation as a whole." For our present purposes, however, it is the nature of the planned hospital provision that is of great importance.

The Plan was a comprehensive review of the nation's hospitals. It provided a national design to try and bring about a distribution of beds based on centrally-determined criteria for matching needs to resources rather than on the haphazard inherited pattern. The criteria were in fact a series of estimates of the appropriate ratios of beds to population in major specialities. For acute beds the recommended ratio was 3.3 per 1000 and a thirteen year period was to be allowed to reduce the 1960 figure of 3.9. There were, however, regional variations from 5.6 in Liverpool to 3.0 in East Anglia. Other recommended ratios included 0.58 per 1000 for maternity (actual 0.45), geriatric 1.4 (1.5), mental illness 1.8 (3.3) and mental subnormality 1.3 (1.3). In the main, what was required was fewer beds of a better standard, in the right places and better used.

The Plan envisaged that this better provision would be brought about by the spatial concentration and centralisation of resources. Its basis was, therefore, the district general hospital (DGH) of around 600 to 800 beds and serving a population of 100000 to 150000. The idea of this facility can be traced to the wartime surveys (see Watkin, 1978) and represented, as Klein (1983, 74) has said, "the · child of a marriage between professional aspirations and the new faith in planning: between ... medical expertise and administrative technology." The DGH was to promote equity and efficiency by ensuring uniform standards throughout the NHS; it "offers the most practicable method of placing the full

range of hospital facilities at the disposal of
patients and this consideration far outweighs the
disadvantages of longer travel for some patients and
their visitors" (Ministry of Health, 1962, 6).

It was very much a plan determined by medical
consensus, and demonstrates the professional dominance of doctors. It is their vision of integrated
services that is presented in the DGH - integrated,
curative, sickness services. But the Plan was only
of limited success. The target bed ratios were
reached by 1967 mainly because of population growth
in the south and the transfer of acute beds to the
growing geriatric service. The capital investment
programme did increase from 9.9 per cent of current
health expenditures in 1966 to 12.6 per cent in
1973-4. It fell back to 9.9 per cent in 1974-5 and
7.1 per cent by 1979-60. The revised hospital plan
(Ministry of Health, 1966) was more uncertain than
the original one, emphasising the need for greater
realism and becoming less definite with regard to
the building programme. The programme always remained vulnerable to the economic crises of the late
1960s and early 1970s and the further revision, the
Bonham-Carter Report on the functions of the DGH
(Central Health Services Council, 1969), which advocated larger hospitals and catchment areas, was
never adopted as policy. Cost implications and
concerns about overcentralisation meant that it was
coolly received. Rationalistic, economies of scale,
arguments were beginning to give way to financial,
cost-containment ones. Indeed, although the 'Best-
Buy' design hospitals built to Ministry standards to
contain and actually predict costs established bed
norms of 2 per 1000 rather than 3.3 at Frimley and
Bury St Edmunds (BMJ, 1976), these schemes were
criticised because of detailed changes made after
tender with resultant increases in costs (Watkin,
1978). Further, the 'Best-Buy' hospital was only
adequate if admissions and discharges were effi-
ciently administered, if aftercare was closely
integrated with community services, if simply surgery was carried out on a day basis and if fuller
use was made of outpatient investigation facilities.
Indeed, as Wheeler (1972) found the norm for Bury
was higher than the 1.92 per 1000 already in operation. Admissions could be increased if length of
stays were shortened. And average length of stays
had shortened for non-psychiatric specialities from
20.1 days in 1959 to 13.3 in 1974 (DHS, 1977a).
Herein lies the nub of the problem of planning that
used bed norms as the criteria of provision. It is

not the number of beds that is important but their
actual use. The number of beds fell 14 per cent
between 1965 and 1977 in England, mainly due to a
substantial reduction in psychiatric beds, a decline
in beds in medical departments and changes in the
treatment of infectious diseases and diseases of the
chest (OHE, 1979). In the same period the number of
in-patients treated rose by 16 per cent.
    National bed norms cannot take into account
local variations in needs. As Klein (1983) says,
there exists an indefinite diversity that helped
subvert the Plan. No two populations are exactly
the same and no two consultants practise the same
kind of medicine. National norms inevitably have to
adapt to local circumstances. Further, subversion
can be attributed to 'the principle of infinite
indeterminancy' in that changes in population structure, in childbearing proclivities and in medical
technology cannot be easily predicted. Again
national norms have to be interpreted flexibly. And
the RHBs did enjoy considerable discretion in how
they allocated their money and organised their
services. As we shall see, such discretion was
exercised not only in relation to acute services but
also with regard to services for the mentally ill
and other Cinderella services which became the
subject of DHSS priority documents.
    But although all these factors helped to dissi-
pate the energy and direction of the Hospital Plan,
its main problem was financial. The comment made by
David Owen, Secretary of State for Health
(1974-1977) summarised it thus "the hospital building programme in 1972-3 like so much public expenditure in this country ... was completely out of
control. Even if Britain had been able to sustain
its then rate of economic expansion the forward
planning of hospitals was completely unrealistic."
This comment begs many questions, eg where should
money be spent, who defines 'unrealistic' and in
what terms, but it points the way forward to issues
of cost control dealt with in next section and
implies a particular view of public expenditure in a
capitalist economy.
    Before we examine this question of state expen-
diture, some more specific issues need addressing.
We have concentrated on the hospital as an illustration of coordinated service provision, the watered-
down version of integrated care. Not only does
hospital planning provide a good example of
rational planning and the social engineering
(expert) approach to care but it allows the most

dominant aspect of social expenditure to come to the
fore. Thus, for example, in 1950, 11.6 per cent of
public expenditure was spent on the NHS, of which
54.9 per cent went on hospital services. This
constituted around 6.5 per cent of public expenditure on hospitals which compares with 9.2 per cent
on education and 8.4 per cent on housing. In 1971,
more public expenditure was spent on hospital services (7.1 per cent) than on the housing budget (6.2
per cent).

Comparison of hospital expenditure with community health service and personal social service
expenditure is also instructive. Within the NHS
budget, the ratio of expenditure on hospital to
community health servics was roughly 7 to 1 in the
period 1949-1956. This advantage declined during
the 1960s to roughly 6 to 1, only to increase to
9-10 to 1 in the period after 1969. We should note
that this worsening of the relative position is
caused not only or principally by the hospital
building programme but also by the transfer of
responsibility for certain services from the NHS to
the local authorities. This change of administrative jurisdiction is often seen as a way of better
providing integrated services. The 1974 reorganisation of the NHS transferred some local authority
services back to the health budget and the school
health services, previously administered by the
Department of Education and Science, were made the
responsibility of the NHS. We would assert that
such shifting of administrative boundaries does
little in the way of improving or integrating services. It demonstrates rather the view that bureaucratic intervention may itself 'solve' problems or
integrate services. It is merely akin rearranging
papers into different piles. Further, within public
expenditure, we may note that personal social services increased their share from less than one
hundredth to one fiftieth between 1961 and 1974.
During this time, the share of hospital services
increased from 5.6 to between 6.5 and 7 per cent,
its high point being in 1973 when they accounted for
7.3 per cent of total public expenditure.

The dominance of hospitals may be explained in
terms of the legacy of past provision and the concomitant revenue expenditure implications. Also
significant is the role of the doctor, particularly
the consultant, in continuing to shape health
policy. Indeed while the spatial distribution of
GPs has been equalised to some extent, less progress
has been made with respect to hospital doctors in

general and certain specialities in particular (see
Stevens, 1966). In 1970, the London regions and
Trent were the worst served in terms of hospital
medical consultants. By reorganisation the pattern
has changed so that the London regions (excluding
South West Thames) had gained at the expense of
other areas. By 1977, Trent, South Western, West
Midlands and Mersey were the worst provided (OHE,
1979). Consultants in teaching hospitals are further more likely to hold merit awards than those in
non-teaching insitutions (see Parker, 1975).

This general pattern, however, masks other
important differences, relevant to our theme of
integration. If integration has been translated
into the coordination of specific services in specific institutions by the dominance of hospitals, its
achievement has been made more problematic by dominance in hospitals and health care of particular
kinds of treatment. George and Wilding (1976) note
that in the 20 years after 1948 the ratio of expenditure per patient in acute hospitals and in long
stay chronic hospitals barely changed, although
there were improvements up to 1973. They calculate
that in 1951 the weekly cost of patients in mental
handicap and mental illness hospitals was 21 per
cent of the cost of acute bed patients. In 1973, it
was 37 per cent. Using OPCS data for 1974-5 (DHSS,
1977a), we calculate that for mental handicap
patients the weekly cost had been reduced to 31 per
cent of that of acute hospital patients. In 1974-5,
the weekly costs were £150.68 in acute non-teaching
hospitals, £137.70 in mainly acute, £160,63 in
maternity, £53.80 in mental illness and £48.02 in
mental handicap hospitals (DHSS, 1977a).

The relatively weak position of certain services
was established, these being labelled the
'Cinderella Services', namely those for the mentally
ill, mentally handicapped and the elderly. This
position, deriving from past neglect, lack of medical and political clout and the dominance of acute
hospital medicine, did not exist in relatively
constrained times. NHS expenditure continued to
rise from under 4 per cent of GNP in the 1950s to
over 5 per cent in the late 1970s and just over 6
per cent in the 1980s (OHE, 1984). The amount spent
on each individual constantly rose, so that in
current prices total NHS cost per head rose from £9
in 1949 to £54 in 1973 and £303 in 1984. If we take
1949 to equal 100, then the index of costs per head
rose to 238 and 341 in the respective years. (OHE,
1984). Between 1957 and 1978, GNP rose by 70 per

cent in real terms while NHS total expenditure went
up by 257 per cent (OHE, 1979). By 1978, health
outlays accounted for 5.3 per cent of GNP, although
their 'best' year had been 1975 with 5.47 per cent,
and it was to rise again in the 1980s (OHE, 1984).
The period 1950 to 1975-6, therefore, saw a great
increase in health and social expenditures (see
Klein, 1983; Butler and Vaile, 1984). Not only did
public expenditure increase (see also Abel-Smith,
1970), but living standards rose dramatically
(Toland, 1980). It should have been against this
confident background that the Cinderella services
were improved and more closely integrated with
housing and personal social services. That was not,
however, quite the case.

Again, there is a somewhat independent history
of interest with these services dating back to the
1950s (see Watkin, 1978; Allsop, 1984). Throughout
the official and unofficial documents, there is the
theme of integrated care in the community. Thus in
1956, the Guilleband Report on NHS costs stated that
policy

> should aim at making adequate provision
> wherever possible for the care and treat-
> ment of old people in their own homes. The
> development of the domiciliary services for
> the purpose will be a genuine economy
> measure and also humanitarian measure
> enabling old people to leading the sort of
> life they prefer.

The White Paper on services for the mentally ill
(Great Britain, 1975) also took as its theme a shift
towards care in the community. But as the quotation
from Guilleband indicates, prioritisation and the
integrative function of community care became implicated in the financial crisis facing the British
state in the late 1970s. Even with adequate funding, it was recognised that resources for health
care were finite and that support for the Cinderella
services would have certain implications.

> We must cease in the health service demand-
> ing more for one sector ... without recog-
> nising that if we take more from one sector
> it has to come from another ... we must be
> prepared to say, if we want priority for
> one sector, where the money should come
> from (Owen, 1976, 113).

Interestingly, it is seen as coming from the family,
friends and neighbours of those requiring care. As
Abrams (1977) notes, in the 1950s and 1960s there
was stress on community treatment - day hospitals,

health visitors, psychiatric nurses in either open
or closed settings. During the 197Gs, community
care became to mean care by the community (see
Chapter Gne). In fact, Care in Action (DHSS, 1981)
regarded the informal sector as the primary source
of community care, with the statutory and private
sectors supplementing and supporting this provision.
Even without the crisis in public expenditure, this
would produce conflicts of interest. The state sees
its own care provision as very costly and family or
neighbourhood care as very cheap. For the individuals, the perception is reversed.

But there has been such a crisis with the late
197Gs seeing a period of high inflation, economic
recession and high unemployment (see Gamble and
Walton, 1976). Capitalist solutions to these problems have been emphasised: the control of wages and
prices and reductions in public expenditure. In
this respect, a move to a less expensive health care
delivery is welcomed. Doyal (1979) notes that this
move also coincided with increasing evidence concerning the limitations of curative medicine.
Consultative documents (DHSS, 1976b; 1977b) argued
that the cost-effectiveness of high technology
medicine had to be reassessed. Resources should be
directed towards the more economical primary care
sector, although as we shall see, the 198Gs has seen
this sector under threat as it appears less economical (being demand-based) than the government would
like. Primary care meant community services, especially for the Cinderella grouping but few extra
resources have been made available. Within the NHS,
the priority groups have not received extra resources
as the struggle for funds has led to the continuing
dominance of acute and hospital medicine (see Eyles
and Woods, 1983). Mechanisms for enforcing the
shift in priorities do not exist and policy-making
in health authorities continues to be dominated by
clinical interests and commitments to existing
projects and expenditure (see Brown, 197G; Hunter,
1979; Haywood and Alaszewski, 198G). This contin-
uance was made all the easier by the consultative
document stating that the shift of resources objective was not a specific target to be reached by a
declared date in any locality (DHSS, 1977b).

While there has been some shift towards care in
the community, most of it, as Allsop (1984) percep-
tively notes, occurred before 1979 and the impact of
the documents. Further, the Care in Action and
Patients First (DHSS, 1979) documents formalised the
shift to care by the community. Priority services

are no more. There are now priority groups with the
main priority being a reduction in expenditure on
personal social services and very low growth in the
NHS. And Patrick Jenkin, Secretary of State for
Health (1979-82) said in 1980, "we cannot operate as
if the statutory services are central providers with
a few volunteers ... to back them up. Instead we
should recognise that the informal sector lies at
the centre with statutory services and the voluntary
sector providing expertise and support" (quoted in
Allsop, 1984, 118). Further, responsibility for
devising community care schemes was placed with
local authorities and RHAs and AHAs, with no funds
for this purpose. This devolution of responsibility
for health care was also based with the <u>Prevention
and Health</u> document (DHSS, 1976c), which emphasised
prevention and health education, especially in times
of resource constraint. Falling ill is seen by
selected example to be caused by individual moral
weakness (Doyal, 1979; see also Crawford, 1980;
Eyles and Woods, 1983). And

> the prime responsibility for his own health
> falls on the individual. The role of the
> health professions and of government is
> limited to ensuring that the public have
> access to such knowledge as is availble
> about the importance of personal habit on
> health and at the very least, no obstacles
> are placed in the way of those who decide
> to act on that knowledge (DHSS, 1976c,
> 102-3).

Thus prioritisation and integrated provision
through community care cease to be the means of
equalising and extending health and welfare services. The devices equally lend themselves to be
rationalisation instruments as was also the case
with RAWP (see Chapters One and Two, and below).
With the priorities programme we have gone 'beyond'
the scope of rational planning and social engineering into the realm of non-planning. We have often
averred that we are not constructing well-defined
stages in the development of health and welfare
policies, as such 'overlap' is inevitable. We have
indeed extended our discussion of this programme to
illustrate the continual interpenetration of ideas
and practices. Just as the social engineering
solution did not simply begin in 1945 nor did cost
containment emerge just from the IMF crisis of 1976
or the Conservative electoral victory in 1979. All
these events of course led to the accentuation of
certain trends and the suppression of others. And

before we turn directly to our final example of an
integrated approach, we wish to examine briefly some
of the issues that led to disaffection from the
social engineering strategy. While we must constantly bear in mind the economic pressures that
fuelled restraint and reduction in public expenditure, we must also examine the spread and efficiency
of government provision of services in the post-1945
period up to the mid-1970s.

We have already noted the tremendous growth in
social expenditure over this period. If we look at
this spending as a percentage of GNP at factor cost,
we may note that from 1951 to 1975 health and welfare expenditure increased from 4.5 to 7.1 per cent,
education 3.2 to 7.6 per cent, housing 3.1 to 4.6
and social security from 5.3 to 9.5 per cent. All
social services spending increased in the same
period from 16.1 to 28.8 per cent (see Peacock and
Wiseman, 1966; Gough, 1979). In terms of public
expenditure, in 1951 health and personal social
services spending accounted for 10.1 per cent of all
government outgoings. This increased to 11.5 per
cent in 1974. In the same period, education spending increased from 6.8 to 11.7 per cent, housing 6.9
to 9.5 per cent and social security from 11.8 to
16.4 per cent (CSO, 1975). It should be noted that
the increase in transfer payments (the transfer of
purchasing power from one group to another - pensions, unemployment and supplementary benefits)
increased more markedly than resource expenditure
(ie that directly consuming labour, energy, buildings and other inputs (see Gough, 1979). The relative significance of health, education and housing
expenditure declined as that of social security
increased. We may note, therefore, that growing
social expenditure is not simply fuelled by the
provision of new and improved services. Growing
social 'needs' (ie greater unemployment increases
the need for such benefit), population changes (ie
more elderly people increases the pension budget) as
well a rising relative costs (ie social services are
labour-intensive and hence productivity gains are
harder to achieve than in industry) generate greater
expenditure. Gough (1979) calculates that if we
take these factors into account the 183 per cent
real increase in personal social services expenditure between 1965 and 1975 falls to a 94 per cent
increase, NHS from 70 per cent to 27 per cent and
education 101 to 29 per cent: still significant if
less dramatic rises.

New and improved services were, however, provi-

ded on a piecemeal basis within the desire of providing a minimum standard and quality of living for
all. The integrating factor of universal benefit
elided to minimum benefits, provided by different
government departments and agencies and requiring
more and more elaborate rules of eligibility.
Bureaucratic intervention appeared to multiply and
means-testing - the basis of selectivity and
anathema of universality - was enforced. Different
entitlements for specific population groups
increased and the structure of health and welfare
services became more complex with, for example,
disability, attendance, and mobility allowances;
special and discretionary payments; family income
supplement; and rent and rate rebates as well as
unemployment, housing and child benefits and retirement pensions. The complexity of the structure
began, it seemed, to defeat its purpose - the
Beveridge ideal of providing a universal, minimum
standard for all. To illustrate this we employ the
analogy of Aitken (The Guardian 6 May 1985) who
compares the labyrinthine complexity of the welfare
state to the behaviour of a person trying to work a
shower with a faulty thermostat control. Not only
is there a time-lag on the lever so the person is
alternatively frozen or scalded by the water, but
new taps have been added piecemeal to try to improve
the situation. There are therefore more controls to
turn frenetically to get delivery right. And more
outlets mean that the pipe from the mains is not
delivering enough water to keep the whole shower
running. But to anticipate, we would argue that
delivery is not improved by ripping out the whole
shower and hence removing the principle of minimum
provision as of universal right.

The implied consequences of such a system are,
however, the involvement of the state in more and
more areas of social life - what Mishra (1984) calls
'government overload' - and a view of the state as a
dominant and stifling force. King (1975, 296)
argues that in the 1970s "governments have tried to
play God. They have failed. But they go on trying.
How can they be made to stop?" One answer to that
question in terms of removing state interference and
inefficiency on the grounds of improviding individuals freedom of choice will be examined in the next
section. This answer may also be seen as an attempt
to change economic practices. Indeed an academic
theory of the state - corporatism - of the 1970s
fits well with the political analysis of the new
Right. As Winkler (1976, 103) writes "corporatism

is an economic system in which the state directs and
controls predominantly privately-owned business
according to four principles: unity, order,
nationalism and success." (The state became directive rather than just facilitative. This view is
best summed up by Pahl (1977, 161):

it could certainly be argued until fairly
recently that the state was subordinating
its intervention to the interests of private capital. however, there comes a point
when the continuing and expanding role of
the state reaches a level where its power
to control investment, knowledge and the
allocation of services and facilities gives
it an autonomy which enables it to pass
beyond its previous subservient and facilitative role. The state manages everyday
life less for the support of private capital and more for the independent purposes
of the state ... Basically the argument is
that Britain can be best understood as a
corporist state.

Corporatism represents a qualitative transformation
of state activities, not necessarily to challenge
private capital but to create "an economic system of
private ownership and state control" (Winkler, 1977,
46). In our view, corporatism seems very much a
creature of its time, representing the apparent
increase in state activities and direction in the
1970s, these features reaching their zeniths in the
late 1970s under Labour and being used to advantage
by Conservatives in their propaganda. And as
Middlemas (1979) points out, the lessons of history
should be that the state is never as directive or
forceful as it might wish to be. The support of
unions and employers was necessary to achieve even
war aims while the state's definition of national
interest may be abrogated where its implementation
would have conflicted with established power relations and interests. To achieve its ends, the state
may even have to bring some powerful interests into
government, eg Labour and the trade unions. The
strengths of the Conservatives and the new Right can
in fact be found in their denial of the role of the
state and their retention of its power of partiality. The redistributive corporatism of the first
30 years of the post-1945 period is rejected to be
replaced not simply by the market as the mechanism
for allocation but by, to use a clumsy term which
gets the balance in the right order, corporate
marketism. As we shall see, the power to undo and

do nothing is equally as great as that to do.
So-called government inaction, the loosening of
control and the reduction of endeavour are just as
directive as their opposites.

Cost Containment and Its Consequences: Financial-
Moral Integration

The changing views on the state as well as economic
crisis contributed to changing attitudes towards the
role of the state and the level of state expenditure.   Economic crisis has been a recurrent feature
of British economic and political life throughout
most of the 1970s and 1980s.   Its management has
been attempted by both Labour and Conservative
administrations, this management usually taking the
form of capitalist solutions to capitalist problems.
As the then Chancellor of the Exchequer, Denis
Healey said in 1976

> Although I have cut expenditure in many
> social fields, I have been increasing
> expenditure in the business field.   The
> relief I have given in tax concessions on
> stock appreciation is higher than any
> relief afforded to business anywhere else
> in the world during a period of inflation
> (quoted in Widgery, 1979, 132).

Britain's position as a low-waged society with lower
'social protection' expenditure per head than its
West European neighbours was being compounded.
These changes were carried out as crisis management
with advocacy reinforcing the effects of reductions
in social expenditure.   As Barbara Castle (1980,
359-60), Secretary of State for Social Services
(1974-6) commented:

> Ministers were never given the chance of
> discussing priorities or overall economic
> strategy.   Instead we were faced with ad
> hoc demands from the Chancellor from time
> to time, pleading sudden crisis or neces-
> sity.   How could I get my health authori-
> ties to plan the NHS properly when the
> capital allocations were abruptly changed?

In this context, marxist and neo-marxist analyses of
the state saw its fiscal crisis being resolved by
its support for the interests of capital in ensuring
the conditions that allowed for the continued accumulation of capital (see O'Connor, 1973;   Miliband,
1977).   This is not the place to replay this debate
in the context of the welfare state (see the discussion in Eyles and Woods, 1983).   Suffice it to note

that even the sophisticated analyses of the state
which highlighted the tension between increasing
support for capital and the increase need to legiti-
mate this partiality (Habermas, 1976; Offe, 1974;
1964) appeared to assume away the significance of
the existence of the welfare state. Its provision
was seen simply as part of the state legitimising
function and by implication a dimension of capitalist domination.

The Left's response to economic crisis has until
quite recently been to advocate class politics with
no real mention of a social programme or to make
such a programme subservient to economic strategy
(see Chapter Two). The recent advocacy (April,
1965) of a social programme of welfare provision and
income redistribution based on significant changes
to the tax system, particularly the abolition of
mortgage interest tax relief in favour of a system
of family-related housing benefit, was shelved by
the Labour leadership without discussion, it being
seen as a 'vote-loser'. The new Right's answer has
been far more appealing to many individuals. In
essence, the legitimation crisis was assumed away,
by their appeals including nearly everyone in the
partial interests and successfully equating capitalist interests with individual interests. Its
success was found in appeal to self-interest, traditional values, the family and economic regeneration.
Its effects on health and welfare services took some
little time to be felt.

These appeals could be found in Conservative
writings dating from the mid-1970s. They emphasise
the integrative principles of policy, namely the
financial (in terms of efficiency, cost containment
and self-provision) and the moral (individual
'freedom', self-discipline, duty and (again) self-
provision). Thus Jordan (1976, 137-139) quotes from
speeches by Sir Keith Joseph:
> ... the only real lasting help we can give
> to the poor is helping them to help them-
> selves; to do the opposite, to create more
> dependence, is to destroy them morally,
> while throwing an unfair burden on society
> (1974).

This burden has not in fact removed social problems
because "in spite of long periods of full employment
and relative prosperity and the improvement in
community services since the Second World War,
deprivation and problems of maladjustment so con-
spicuously persist" (1972). This public burden view
of welfare was to become extremely influential in

the late 1970s and affected Labour's welfare strategies. Public services were seen as a drain on the
productive sector (Bacon and Eltis, 1976) - a financial as well as moral burden. To return to Sir
Keith:

> Parents are being divested of their duty to
> provide for their family economically, of
> their responsibility for education, health,
> upbringing, morality, advice and guidance,
> of saving for old age, for housing. When
> you take responsibility away from people,
> you make them irresponsible. Hand in hand
> with this, you break down traditional
> morals (1974).

Jordan persuasively juxtaposes views from the 1834
Poor Law Report with this speech to demonstrate that
the same view of a seduced, debilitated working-
class is being presented. The moral necessities
(integrative elements) of cost containment appear,
therefore, as an attempt to instigate a new New Poor
Law.

The appeal to individualism may be seen through
what Mishra (1984, 83) calls "the contradictory
nature of social democracy". While the bureaucratic
intervention of the consensual post-1945 social
democracy represented working-class (or 'ordinary')
people's hopes and interests, it did so within the
confines of capitalism. It had to 'discipline'
workers and their organisations and appeal to
national rather than class interests. These problems are particularly exploitable in times of economic crisis. Further, extension of state services
occurred without any attempt to mobilise democratic
support for such actions. "The state is increasingly encountered and experienced by ordinary working people as, indeed, not a beneficiary but a
powerful bureaucratic imposition" (Hall, 1979, 18).
This imposition on 'individual freedom' could be
exploited by the new Right particularly in the
context of government departments and local authorities seeming to have no way of assessing the impacts
of their expenditures. Commenting on social services planning, the House of Commons Social Services
Committee stated in 1980

> rather than deciding upon an overall
> strategy and then adjusting the various
> elements of the strategy accordingly,
> policy is made by taking decisions about
> specific items (according to whatever
> criteria may be in use) and then having a
> retrospective look to see what their com-

bined effect turned out to be (quoted in
    Walker, 1964, 161).
Such muddle could be and was accused of generating
wasteful expenditure so it was not services that
were to be cut but this 'surplus' so that resources
could be freed for productive investment and taxes
cut to generate entrepreneurial spirit and enhance
individual choice.

Such were the persuasive appeals in 1979. The
public sector was seen as an unproductive burden on
individual taxpayers and wealth creation. The real
cost of public services must be kept down and any
rise prevented. Such aims would be achieved by
establishing price stability through control of the
money supply with the main role of government being
to ensure the conditions for the proper functioning
of markets (see Gamble, 1980). The first
Conservative budget (1979) had in fact stated: "it
is crucially important to re-establish sound money.
We intend to achieve this through firm discipline
and fiscal policies consistent with that, including
strict control over public expenditure" (quoted in
Walker, 1984, 114). The integrative principles of
government action are encapsulated in the following
statement on strategic guidelines which are

    first, the strengthening of incentives,
    particularly through tax cuts, allowing
    people to keep more of their earnings in
    their own hands, so that hard work, ability
    and success are rewarded; second, greater
    freedom of choice by reducing the state's
    role and enlarging that of the individual;
    third, the reduction of the borrowing
    requirement of the public sector to a level
    which leaves room for the rest of the
    economy to prosper; and fourth, through
    firm monetary and fiscal discipline, bring-
    ing inflation under control and ensuring
    that those taking part in collective bar-
    gaining are obliged to live with the conse-
    quences of their actions (H.M. Treasury,
    1979, 1).
The language employed is illuminating.

How well have these policies worked? Have costs
been contained? If we examine the latest figures
available (H.M Treasury, 1985), we note that public
expenditure in real terms has increased by 8 per
cent between the 1979-80 outturn and the 1985-86
plans. Further, there has been a transfer of expenditure from some programmes to others. Over the
same period, social security (30 per cent), agricul-

ture (26 per cent), law and order (27 per cent),
defence (23 per cent) and health and social services
16.5 per cent) have 'gained' in real and comparative
terms. Of other identifiable programmes, Northern
Ireland (5.5 per cent) and education (4 per cent)
have 'gained' in real terms but 'lost' comparatively, leaving the following as 'double-losers':
Scotland (-1 per cent), arts and libraries (-1 per
cent), Wales (-2.5), transport (-13 per cent),
overseas aid (-17 per cent), environmental services
(-18 per cent), industry, energy, trade and employment strategies (-26 per cent) and housing (-68 per
cent). Public expenditure has not so much been
reduced but transferred and concentrated with the
significant 'gainers', with the exception of agri-
culture, being 'protection services' of both a
coercive and a social kind. Government has not so
much removed itself from the lives of its citizens
but become far more partial in its allocations, with
its main commitments being to defence and public
safety and to the burgeoning social security bill
driven remorselessly upwards by an ageing population's claim to pensions and by deindustrialisation
and concomitant increases in unemployment (which
themselves have health expenditure implications -
see Chapter Three and Four).
     It seems though that health and personal social
services have been protected. But if we employ
recent statistics (CSO, 1985) it is possible to make
the following calculations. In current cost terms,
government expenditure on the NHS current account
rose in total 66 per cent between 1979-80 and 1983-4
and 16 per cent between 1981-2 and 1983-4. The
hospital sector, which dominated expenditure in the
bureaucratic interventionist period, enjoyed respective rises of 66 per cent and 3 per cent. If we
take just the general medical services element of
FPS we calculate the respective rises to be 86 per
cent and 21 per cent. If we examine personal social
service current expenditure in the same way, the
rise was 63 per cent between 1979-80 and 1983-4 and
19 per cent between 1981-2 and 1983-4. The respective increases for central government expenditure
were 20 and 10 per cent and for local authorities 64
and 19 per cent. We should also note that the
decimation of the housing programme as shown by the
White Paper on expenditure plans has had a marked
effect on public spending on housing. While there
has been a close-on 400 per cent decline in housing
subsidies to local authorities, there has been a
four-told increase in payments under the rent rebate

scheme and a fivefold increase under the rent allowance scheme. There may well be financial integration at the level of fiscal policy but the consequences of actions in one arena of policy on another seem to have escaped attention. As the House of Commons Social Services Committee was moved to say

> on the basis of the evidence we have heard, we are struck by the apparent lack of strategic policy-making at the DHSS: the failure to examine the overall impact of changes in expenditure levels and changes in the social environment across the various services and programmes for which the Department is responsible....At this point, the Committee wishes to record its disappointment - and dismay - at the continuing failure of the DHSS to adopt a coherent policy strategy across the administrative boundaries of individual services and programmes (quoted in Walker, 1984, 149).

While this comment also applies to the DHSS in relation to other departments, the lack of coherence means that cost containment has the impact of passing patients and clients 'down the line', until they become someone's statutory responsibility, be it supplementary benefit offices or personal social services. Cost containment has unintended (and costly) consquences. This may be exemplifed by the massive increase in the use of bed and breakfast accommodation to house homeless and often ill people who become the responsibility of personal social services. This uptake is the unintended effect of cuts in the housing programme and appears to have a marked spatial concentration in large cities, particularly in inner London. Of course, such individuals can only be 'housed' in areas where there are numerous small hotels and boarding houses, which are located in London mainly in and on the fringes of the central area. At least 18 boroughs are searching for bed and breakfast accommodation, often outside their own boundaries. There are, however, different DHSS rates for those resident in different parts of London with the ceiling in outer districts being two-thirds of that in inner areas (see Conway & Kemp, 1985). This means that boroughs have to pay any difference between accommodation costs and DHSS allowances. These differences may be marked as 'hotel' owners see competing local authorities willing to pay high costs to house their homeless. Owners may collude to increase prices so that rates of return are extremely high. One 150-bed central London hotel is estimated to take £0.75 million a year. Such accommodation is the preserve of those with other disadvantages, particularly the unemployed, single parent families and ethnic minorities. These problems are compounded as overcrowded, often insanitary conditions lead often to behavioural and health difficulties. Intestinal disorders and cases of malnutrition have been found. Looking for employment becomes difficult from such a base which is also likely to lead to an increase in stress-related disorders. Thus the costs of the housing programme are contained at the price of increasing social security, social service and local authority expenditure and of individual hardship and ill-health.

While these unintended consequences are extremely important, we wish to turn to more general implications of the present financial-moral integrative principles. In general terms, cost containment has been achieved through cash limiting expenditure. Cash limits were first introduced in 1975 under Labour and were extended to more arenas of policy in 1976 because of the 'economic or IMF crisis'. Under the Conservatives, cash totals rather than survey prices were used in expenditure plans, a change that intensified control of spending. With cash limits, it is primarily the Treasury and its rather narrow economic criteria that dominate debate. Thus, in general terms

Despite protestations from some government departments, like the DHSS, that their planning is 'needs'-conscious, the fact is that the exigencies of the economy, as decided by the Treasury, have led to the adoption of public expenditure control as the dominant form of planning approved by the Cabinet and imposed by Whitehall (Townsend, 1980, 8-9).

These narrow economic criteria emphasise the importance of the private sector in a capitalist society, like Britain - a view that dominated Treasury thinking at the time of Beveridge (Walker, 1984) - and posit a skewed view of public expenditure in that tax expenditures are excluded from consideration. Without these, true integration of social, let alone social & economic, policy is impossible.

As a form of planning, cash limiting is cost-dominated. This means that it emphasises the resource implications of policies rather than their benefits. It does not generate any rationale for

the alternative use of resources. A major reason
for this is that a constrained cash limit forces the
agency or department concerned to define clearly its
own sphere of statutory responsibilities. If its
budget is limited, it is unlikely to provide that
which it has no legal or overtly moral need to
provide. This position may well be assisted by
appeals and exhortations that communities and families should care for themselves and by statements
which argue that the state has a limited role and
responsibility (see above). The move towards cost
containment may be seen in this context to have two
disintegrating tendencies. First, it points the way
to individual and local care provision. We shall
treat individual care as private care more fully
below and state simply that such provision leads to
the plurality of organisations rather than their
integration. As Patrick Jenkin said

> I believe it is wrong to treat the NHS as
> though it were or could be a single giant
> integrated system, rather we must try to
> see it as a whole series of local health
> services serving local communities and
> managed by local people (DHSS, 1979).

Just how this might be achieved and funded is not
stated, although the role of the private sector
must be seen as significant. Secondly, joint care
planning, instigated in 1977 for improving service
delivery to clients of the NHS and personal social
services, becomes a financially difficult exercise.
Without major financial shifts in resource allocation (difficult to achieve under cash limiting)
joint planning remains largely a bureaucratic exercise, although it attempts at co-ordination and
collaboration should not be discounted. In 1963-4,
0.6 per cent of the NHS budget and 4.5 per cent of
the personal social service budget were given over
to such collaborative ventures (Walker, 1984).
While uptake of allocation is extremely high, differences in accountability and wide local discretion
as well as small size and short duration of the programme (see Booth, 1981; Glennester, 1983) militate
against its integrative effects (see Chapter Six).

Cash limits began under Labour and this particular response to economic crisis led to public
expenditure reductions in the period 1976 to 1979.
We have already noted how NHS expenditure as a
percentage of GNP peaked in 1975 at 5.47, falling to
5.32 in 1978 (OHE, 1964). It rose to 6.26 per cent
in 1961 as a result of pay settlements agreed under
Labour and paid under the Conservatives. The prog-

ramme of closing small, specialist hospitals and
reducing the increases in staff levels also began in
the mid-1970s, a partial and polemical account being
provided by Widgery (1979) and statistical evidence
in DHSS (1982). What was also initiated under
Labour was a device we have discussed before, the
formula of RAWP. The initial aim of RAWP was to
restructure the delivery of certain types of health
care in a redistributive manner. But the inclusion
of RAWP in cash planning has meant that real reductions in expenditure and staffing have occurred
within some regions. Particularly adversely
affected have been the four Thames regions. The use
of sub-regional RAWP formulae has meant that inner
districts of these regions have been and are to be
seriously constrained. Thus the Thames regions lose
out in the national restructuring of funds for
hospital expenditure and the inner districts,
because of their declining populations and 'over-
provision of hospital beds, lose out in the regional
restructuring.
    The effects of cost containment on the RHA have
been documented by Mohan and Woods (1985). For the
English regions, expenditure was reduced in real
terms in the financial years 1983-4 and 1984-5 and
there were contained attempts to reallocate
resources from the south east to the north and west.
Table 5.3 illustrates these themes as well as the
effects of changing policy. In January 1985, an
additional 1.2 per cent revenue was allocated to the
NHS and was differentially distributed to the
regions (col 3) on the basis of their distance from
revenue targets (col 2). Of the 1.2 per cent, 0.5
per cent was to come from existing budgets, via
efficiency savings, which were pooled and
reallocated. But as columns 4 and 5 show, the
expected rate of growth was reduced, because of
general financial stringency. This meant that the
Thames regions could expect to receive less in
absolute terms each year. With N E Thames, over £10
millions were lost, even after a £2 million transfer
from the regional capital budget. As Table 5.4
shows, there cuts varied from £300,000 in Redbridge,
West Essex and Southend to £1 million and over in
the inner districts of Bloomsbury and City and
Hackney. Table 5.3 also shows the attempts to cut
manpower (col 7) with areas of high unemployment
(Mersey, Northwest) affected as well as the Thames
regions, this being an important consideration given
the importance of NHS employment. Further we may
point to Lambeth district which must reduce expendi-

Table 5.3: NHS financial and manpower changes 1983-5

(1) Regional health authority	(2) Distance from revenue target as a percent of 1962/63 allocations	(3) 1983/4 services development addition (January as a percent of 1962/83 allocation	(4) 1983/84 service development addition/reduction (July) as a percent allocation	(5) Annual growth/reduction (per cent) for ten years until 1993/94	(6) 1984/85 service development addition as a per cent of 63/84 (July) in real terms	(7) Manpower changes between 1963 and 1964
East Anglian	- 9.46	+2.91	+1.96	+1.6	+1.9	+ 374
Trent	- 7.55	+2.41	+1.40	+1.1	+1.6	+ 520
Wessex	- 7.06	+2.11	+1.16	+1.4	+1.8	+ 40
South Western	- 6.04	+1.65	+0.65	+1.3	+1.6	- 124
Yorkshire	- 5.94	+1.61	+0.61	+0.6	+1.3	- 264
North Western	- 5.33	+1.25	+0.25	+0.4	+1.3	- 562
West Midlands	- 5.13	+1.31	+0.31	+1.0	+1.4	- 140
Northern	- 4.96	+1.21	+0.21	+0.5	+1.4	- 186
Mersey	- 2.60	+1.11	+0.11	+0.2	+0.6	- 506
Oxford	- 1.41	+1.45	+0.45	+1.4	+1.7	+ 229
South East Thames	+ 5.71	+0.35	-0.64	-0.3	0.0	-1081
South West Thames	+ 6.06	+0.35	-0.64	-0.3	0.0	-736
North East Thames	+ 8.04	+0.32	-0.68	-0.3	0.0	-1200
North West Thames	+10.43	+0.31	-0.68	-0.5	0.0	-1000
Total		+1.22	+0.21	+0.5	+1.0	-4037

Source: Mohan & Woods (1965)

Table 5.4:   The effect of cuts: district revenue allocations
             1963-1964 in the North East Thames Regional Health
             Authority

Less

Health District (1)	Recurring Base £000 1.4.63 (2)	Efficiency Cut £000 · (3)	Lawson Cut £000 (4)	Total £000 (5)
Basildon & Thurrock	42,122.6	120.7	302.6	423.3
Mid Essex	37,905.6	111.0	260.7	371.7
North East Essex	47,968.0	137.4	346.6	484.0
West Essex	26,454.1	135.8	203.4	339.2
Southend	34,563.7	98.1	249.3	347.4
Barking, Havering & Brentwood	61,185.0	293.4	435.0	728.4
Hampstead	59,177.0	283.8	421.0	704.8
Bloomsbury	107,156.7	717.3	758.0	1,475.3
Islington	42,835.4	287.6	303.4	591.0
City & Hackney	72,388.9	486.0	513.2	999.2
Newham	31.246.7	209.6	221.4	431.0
Tower Hamlets	61,938.8	415.9	439.3	855.2
Enfield	30,178.4	135.6	215.1	350.7
Haringey	34.909.5	154.0	247.7	401.7
Redbridge	31.461.6	100.5	224.0	324.5
Waltham Forest	54.940.1	263.5	390.1	653.6
RHA Services	43.310.9	217.1	307.2	524.3
Totals	821,761.4	4,167.3	5,838.0	10,005.3

Source: Mohan & Woods (1985)

Lure so that the under-resourced districts in Kent
of S E Thames RHA may benefit. But if different
criteria were used in RAWP or sub-regional RAWP cash
limiting and restructuring would take different
forms. If, for example, the nature as well as
quantity of cross-boundary flows and the characteristics of resident populations and their health
requirements were more fully taken into account,
different restructuring would be suggested. Cash
limiting is a crude a device which cannot take on
board nuances of social differentiation or geographical scale. These require different conceptions
of justice - a more interventionist, need-oriented
view - to be treated. Indeed, overall RAWP policy
shows planned growth varying from 1.9 per cent in
East Anglia to zero in the Thames regions. But even
this is undermined by unrealistic assumptions concerning wage and price inflation. The containment
of hospital costs through RAWP has accelerated
therefore, hospital rationalisation and the concentration of services on fewer sites. From 1979-83
there were 37 hospital closures and 15 partial
closures in the Thames regions. N E Thames reduced
its bed capacity by 12.9 per cent over this period.
Public hospital distribution is thus becoming
restructured and more geographically uneven with
concomitant accessibility problems.
    Cash limits cannot be easily applied in the
demand-determined arenas of public expenditure,
including FPS and social security, the latter containing payments made by social service departments.
We noted above how real growth in the health budget
can be largely accounted for by FPS. We used calculations for general medical services to illustrate
this point. Using the same source (CSO, 1985), the
increase in expenditure on pharmaceutical services
(excluding patient contributions) was 74 per cent
between 1979-80 and 1983-4 and 27 per cent between
1981-2 and 1983-4 (total current expenditure on the
NHS increased respectively 68 and 16 per cent).
Doctors and the dispensing of their prescriptions
are a significant proportion of increased health
expenditure. This increase may reflect additional
demands placed on FPS by the stresses and strains
suffered by particular individuals and groups (see
Chapters 3 and 4). It means that other methods of
cost containment may have to be tried, especially
after the virtual withdrawal of government directives not to prescribe certain expensive prepara-
tions in the face of clinical autonomy. It is in
these other methods that the conjoining of financial

moral criteria may be seen.

Increasingly, appeal is made to self-provision of care. With regards to GP services, it is likely that some thought will be given to some form of 'privatisation', including possibly the introduction of health maintenance organisations (HMOs). The basis of an HMO is a contract between GP and patient by which the former provides the latter with a comprehensive care package. The GP provides primary care and buys in hospital care from the state or private sector on the best possible terms. GP income would depend not on capitation payments, fees and expenses but on how many patients that s/he could attract and her/his ability to control costs. Patients would be provided with vouchers which would entitle them to so much care and so many prescriptions a year. The elderly and long-term sick would obtain additional vouchers. HMOs are meant to contain costs by abolishing service payments to GPs and making them responsible for costs. While a private care organisation exists at Harrow in north London, British thinking has been mainly based on American practice where HMOs have cut health care costs, although they are still twice as expensive as NHS treatments. HMOs have raised the ire of the British medical profession, so implementation is perhaps unlikely. They also introduce the possibility of links between hospitals and GPs which may not be in patients' interests and challenge one of the foundational principles of the NHS, equality of access irrespective of income.

While such privatisation is still being discussed, self-care through subscription to private insurance schemes to cover GP and hospital and specialist treatment has grown substantially since 1979. There were then less than 2.5 million people covered by such insurance. Now there are around 4.5 million (Manwaring and Sigler, 1985). There were in 1984 some 190 private hospitals in Great Britain, containing around 6,500 beds. This has now risen to over 10,000 (Day & Klein, 1985) Private health care encapsulates the integrative principles of cost containment and individual responsibility. Kenneth Clarke (Minister of Health) stated in 1984 that "health authorities should collaborate with the private sector to avoid wasteful duplication of facilities" while the 1983 Conservative party manifesto averred: "Conservatives reject Labour's contention that the state can and should do everything. We welcome the growth in private health insurance in recent years. We shall continue to encourage this

valuable supplement to state care" (quoted in
Manwaring and Sigler, 41,40). And so they have with
interesting social and geographical consquences.

It is interesting to note how overtly geographical concerns re-emerge with the consideration
of private health care. When we examined bureaucratic intervention, these concerns became less
explicit because the principles of integration
suggested universality and equivalence of provision
between groups and across space on the basis of
centrally established 'norms'. That these norms and
guidelines, that central direction and broadly
consensual planning had not achieved equality of
access was demonstrated by the early pronouncements
of RAWP and the continued existence of over- and
under-doctored areas. But it is provision based on
cost containment and through private insurance that
has brought the geography of care to the fore. The
private provision of health care is a geographically
uneven development. In saying that, we must of
course note the strong relationship between private
care and social class which in part subsumes the
geographical relationship. Those in professional
occupations are eleven times more likely to be
covered by private insurance than those in manual
jobs (OPCS, 1984). But as Day & Klein (1985) point
out even when allowance is made for the social
composition of the population, coverage is 50 per
cent higher than would be expected in Greater London
and 50 per cent lower in the north. They tenta-
tively suggest that the reason for this is that
certain consumers see certain advantages in private
treatment. They do not reject the NHS. They prefer
to use the private sector to their own individual
advantage. Day & Klein (1985, 1292) argue then that
"it may be that the greater consumption and concentration of private health care in Greater London,
and parts of southern England, reflects the fact
that this is where the mobile, consumer oriented
middle classes are concentrated, with somewhat
different cultural values than their social equivalents in the north." Individualist responses as
opposed to collectivist ones may, therefore, dominate, resulting in a greater propensity to insure
and use private health care.

Insurance coverage varies socially and spatially
(OPCS, 1984; Mohan, 1985); 23 per cent of those in
professional occupations were covered by health
insurance compared with 2 per cent of those in semi-
and un-skilled manual jobs. While 15 per cent of
the population of S W Thames RHA is covered, only 4

per cent of those in the Northern region area.
Overall, coverage varies from close to zero in inner
London to nearly 20 per cent in parts of
Buckinghamshire. But insurance coverage is only one
of the factors influencing the distribution of
private hospital facilities. Indeed, some of these
facilities have been developed by non-profit making
organisations and these show a fairly dispersed
pattern (see Figure 5.1). Profitability is not a
criterion for such organisations as Nuffield Hospitals and their small units may be found in Hereford,
Chester, Shrewsbury, Exeter and Lancaster. Fig 5.1
also shows the location of profit-making institutions. This, however, masks the dominance of London
as the prime locus of private hospitals (see Fig
5.2). N W Thames has 39.8 private beds per 100,000
population and N E Thames 32.1 compared with the
Northern RHA's 2.1 (see Griffiths & Rayner, 1985).
While Greater London possesses over half the private
hospital beds, the two North Thames regions contain
about one-third of the national total with a particular concentration in central London eg Bloomsbury,
Harley Street, Victoria.
    The proximity of consultants and the initial
effect of private consultation rooms are important
locational determinants. Locations near NHS facilities are highly prized because consultants can
easily work in both state and private hospitals.
The private Cromwell hospital in Victoria has four
teaching hospitals within a 4 kilometre radius
(Mohan, 1984b). As well as profitability and consultant availability, Mohan (1985) also points to
the role of capital put up by local consultants and
business people, the importance of green-field sites
to create an attractive environment for patients and
town and country planning regulations are being
influential in determining the distribution of
private hospitals.
    The demand for such facilities is not, however,
infinite. Figure 5.3 shows the notification of
intent to apply for planning permission for private
hospital development in the period 1980 to 1983. It
must be debatable how many of these schemes will
come to fruition as oversupply of private beds may
be the case in at least London, Birmingham, Glasgow
and Bristol. Expansion of the number of beds has
reduced occupancy rates. In 1984 average bed
occupancy was 65 per cent, compared with the break
even point of 70 per cent (Day & Klein, 1985). This
means that cost containment, against, a backdrop of
a sluggish increase in insurance coverage (3 to 4

Figure 5.1 Private hospitals in Britain

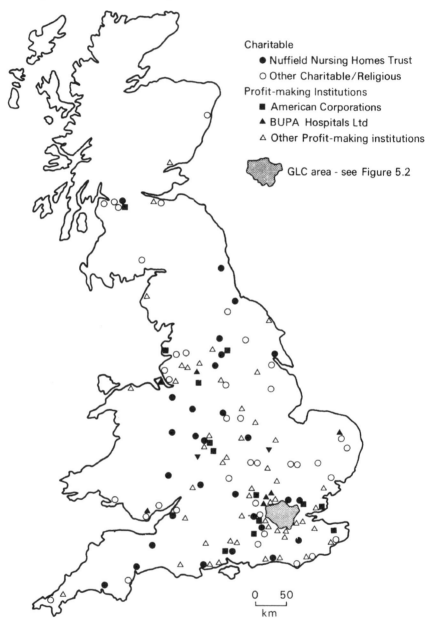

Charitable
- ● Nuffield Nursing Homes Trust
- ○ Other Charitable/Religious

Profit-making Institutions
- ■ American Corporations
- ▲ BUPA Hospitals Ltd
- △ Other Profit-making institutions

GLC area - see Figure 5.2

0   50
km

Source: Mohan (1985)

Figure 5.2 Location of capacity of private hospitals providing acute surgical facilities in the Greater London Council (G L C) area in 1983

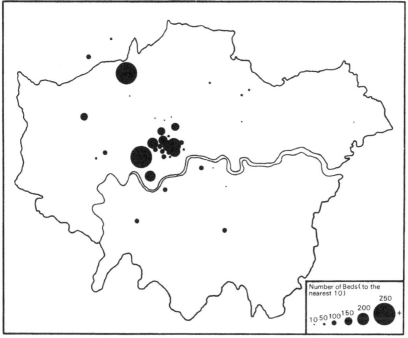

Number of Beds (to the nearest 10)

250

10 50 100 150 200 +

Source: Mohan and Woods (1985)

197

**Figure 5.3 Notifications to the Secretary of State for Social
Services of intent to apply for planning permission
for private hospital development, 1980-3**

Source: Mohan (1984b)

per cent in 1965) and rapidly rising subscriptions
(up 61 per cent between 1981 and 1983 compared with
a 14 per cent rise in the retail price index) (The
Financial Times 22 January, 1986), is the order of
the day in the private sector. To contain costs,
direct payment by insurers has been introduced and
there have been advances into less costly areas of
care, notably geriatric and psychiatric care. Thus
Kneesworth House near Cambridge for mentally dis-
ordered adults offers a variety of therapies for
fees of £650 per week, while it has been estimated
that 5 per cent of the elderly could afford over
£200 per week for high quality residential care (The
Financial Times, 22 January, 1986). Further, while
it is unlikely that the Conservative government will
intervene to regulate the provision of private care,
joint ventures with NHS facilities are likely to
increase. In fact, more than 40 per cent of the
health authorities in England and Wales have some
kind of contractual arrangement already with the
private sector. For example, Portsmouth, Bath,
Blackpool and Newcastle have contracted with private
hospitals to carry out some operations where the NHS
waiting list is too long. Private hospitals carry
out some 10,000 operations a year for the NHS. But
these joint ventures point as much to the circum-
scription of the NHS through strict financial
limits, privatisation of tasks, changes in management structure as to the role of the private sector.
The private sector is likely to serve specific
social groups and geographical areas. Its growth
may thus be self-limiting, although we should note
the great loyalty demonstrated towards the NHS by
most of the British public as a universal service,
'free' when required and distributed according to
need. If private medicine succeeds in chipping away
this loyalty, it may strengthen its own promotion
(see Mooney, 1983; Griffiths & Rayner, 1985) as an
image of a two-tier care system becomes clearer,
namely the private sector for those with sufficient
funds to buy 'choice' or insurance against illness
and the NHS as a safety net for those who cannot
afford this choice. In terms of general distributional policy, the private sector seems to be working against the redistribution of resources for
health care away from the south-east and towards the
north and from central city to outer metropolitan
areas (Mohan & Woods, 1985). But with the private
sector having similar cost problems as the NHS, its
promotion and growth are likely to become increasingly problematic. Its existence does, however,

distort the nature and priorities of the NHS through
providing different opportunities for medical practice. Its problems though lead to the long-term
possibility of integrating private and public provision through modificiations to the RAWP formula to
account for private hospitals as happens in New
South Wales, Australia (Eyles, 1985b) and some
amendments for professional autonomy in HMOs. In
any event, with cost containment being the prime
concerns of policy, the private sector, as a form of
self-care, is and will remain a significant dimension of health care provision.
    Self-provision is also encouraged by the idea of
community care which as well as appearing humane is
also an effective cost containing strategy. The
general social and gender implications of this
strategy were spelt out in the mid-1970s by cam-
paigns against hospital closures in the Islington
area:

    The propaganda campaign for 'community
    care' is very dishonest and mainly aimed at
    getting the public to accept the run down
    of the NHS in the vague belief that other
    facilities exist. The trick is they don't
    and community care means that we will have
    to care for the chronically, but not criti-
    cally, sick young and old in the home.
    This can only be done if more and more
    women stay at home to take on this task.
    We do not say that only women can care for
    the sick but we expect that in practice
    this will be what happens. Without much
    support from either the health service or
    the local council this could well mean a
    reversion to the bad old days of imprison-
    ment in the home for many women (quoted in
    Widgery, 1979, 167-8).
Community care is linked to the prioritisation of
care for particular groups. But these are the
groups that suffered from reductions in local
authority expenditure, for example, less real
expenditure on meals served at day centres or in
people's homes, on local authority residential care
places for the elderly, on holidays for the elderly
and the handicapped and on aids and adaptations for
the physically handicapped (see Manwaring and
Sigler, 1985). Costs have been reduced but other
forms of care, except the informal and virtually
hidden assistance of family and friends, have failed
to materialise. Voluntary organisations are themselves handicapped by the reductions in contribu-

tions from local authorities. And these authorities
cut back where they have no statutory responsibility. The appeal to self-provision and individual
and community responsibility within the context of
cash limits is being accompanied by a reality of
collective irresponsibility as exhortation replaces
cash and commitment. The question who cares, especially for those prioritised as the Cinderella
groups, may be put in a double sense as a question
answered by state, local authority, family or individual and as a perojative statement assuming the
answer no-one.

To illustrate the changing nature of priorisa-
tion further, we may turn to a more geographical
example, the inner city, which has been the priority
area of governments from the late 1960s through to
the present. While we must note the dangers of the
ecological fallacy, the vaguely defined inner city
has become seen as the repository of many social and
health problems. The concentration of these problems may be due partly to the age of their housing
stock, partly to the state of their economic bases
and partly to selective in- and out-migration, with
their losing the young and skilled, retaining the
old and unskilled and gaining overseas immigrants
and their descendants (see Eyles, 1979). Such areas
tend to have higher than average rates of unemployment, housing stress, educational disadvantage, and
psychiatric and health disorders. Interestingly,
such areas have poorer than average primary health
care facilities with a significant proposition of
older, solo-practice GPs (see Chapter Four).

The policies and programmes to tackle the inner
city problem reflect well the changing nature of
policy that we have thus far described. Initially,
much emphasis was placed on housing improvement.
Such improvement is itself vital and may be equated
with the public health measures of the 19th century
for its impact on general quality of life. It does,
however, represent par excellence a social engineering solution in that changes wrought by experts, in
this respect at one remove, their being environmental rather than social changes, were seen as the
way to improvement. Throughout the 1960s and 1970s,
bureaucratic intervention increased in its range and
direction if not in its generosity. The urban aid
programme allowed supplementary funds to be directed
at deprived groups and communities (see Eyles, 1979;
Lawless, 1981; Berthoud et al, 1981). But the
grants provided were small in amount, duration and
scale. Such a parsimonious and piecemeal approach

was seen as increasingly unsuitable as unemployment and inflation rose in the late 1970s. Further, it seemed unlikely that better coordination of the personal social services and the encouragement of mutual aid and self-help would lead to lasting solutions to the inner city's social and economic problems.

A new initiative was launched (DoE, 1977b) which in the light of economic decline producing high rates of unemployment and a mismatch of skills and jobs; the physical dereliction of the areas and the extent of social disadvantage would "give the inner areas an explicit priority in social and economic policy." Funding was increased and in the designated partnership areas e.g. London Docklands, inner Birmingham, Newcastle-Gateshead, joint collaboration between central and local governments was initiated. We may regard this programme as the culmination of the bureaucratic intervention - social engineering period. Integration of policy was seen in terms of a broadly conceived state-provided solution (facilities, jobs, improved infrastructure).

This imposed, broad-based 'solution' went the way of all such endeavours in 1979, although it must be said that even up to this date little extra cash was made available. With Conservative strategy, individual responsibility and self-provision were seen as the necessary steps to progress. Government could assist by removing some planning controls and by stopping intervening in the areas, intervention being seen as preventing private enterprise and investment. Enterprise zones (see Butler, 1982) were established in a variety of derelict or problem areas e.g. London Docklands, Corby. Controls were not relaxed as much as originally envisaged but the chairman of London Docklands Development Corporation described the zone as containing "the most benevolent planning and taxation regulations ever dreamed of" (Cross, 1982, 711). Enterprise zones have not created many new jobs or solved many problems (see GLC, 1984). Rather, they have shifted the geographical location of jobs and problems. Firms are encouraged to relocate in them and grants to improve facilities and infrastructure have a noticeably deleterious effect on neighbouring districts (see O'Dowd & Rolston, 1985). They have, by definition, also been concerned chiefly with employment and have done little for service provision, which remains cash-limited. Indeed by creating low-wage districts (and essentially low-wage employment has been attracted while high-wage work has gone to

green-field, often small-town sites), such zones may
indirectly exacerbate other problems such as poor
diet, family poverty and so on. They

cannot combat structural unemployment or
underemployment, and they ignore other
social and economic problems of our cities:
insufficient private and public investment,
inadequate educational opportunities,
concentrations of the elderly and of people
who have never experienced work, a deterio-
rating physical environment. These problems cannot be solved by substituting 'free
enterprise' for government (Mier, 1982,
14).

But it has been that substitution that has run
through this final section. Further, it is a substitution which has been derived from emphasising
cost containment (and the concomitant anti-statism)
and individual responsibility. To alter this pattern, however, requires more than the resubstitution
of government for free enterprise. It entails
consideration of socialist integration (see Chapter
Six).

Conclusion

We shall reserve discussion of socialist integration
until the next chapter. This chapter has already
ranged widely over a great deal of British social
and economic policy. We have though been selective
- a practical as well as substantive necessity. We
have selected to show how policy in the health and
welfare fields has been integrated. This integration is of course often at the level of policy
statement and commitment rather than practice. But
we still wish to argue that policy matters have
tended to see social and health problems or their
solutions in unified ways. The New Poor Law saw the
cause of the problem as moral degeneracy. In some
respects, the solutions - individuals diligence and
thrift with the horrors of the workhouse and its
associated hospital in the background - were also
integrated. But there were other providers of care
(the voluntary and local authority bodies), pressures from reformers and the working class for
change, and reports that gave the lie to the universal cause of poverty. To state the obvious, practice is so much more complicated than theory as
reality interposes. Such must also be the conclusion of the first 30 years of the post-1945 experience. Its universal solution-minimum provision for

all - broke down under the plethora of 'special
cases' or problems requiring attention. The last
ten years have in some ways tried to integrate
problem and solution by the denial of both. A
minimalist solution has been established whereby
cash planning allocates resources to programmes
(after much acrimonious debate at all decision-
making tiers). Backed up by a philosophy that
emphasises individual freedom and provision, the
only alternative to the cost constrained policies is
to provide it oneself. As we shall try and show,
there is another alternative.

This chapter has also been about how resources
have been allocated. We must draw attention to some
of the convergences between this chapter and Chapter
Two. The broad and ill-defined periods that we have
examined have their rough equivalences in terms of
views on need and justice: 'rough' because not only
are concepts of need and justice contestible but
because they are contestible at all times. Different groups possess different conceptions which they
argue for and try to get implemented. Hegemony is
never all-embracing. It contains disjunctures and
contradictions and any society usually possess other
meaning-systems (see Parkin, 1971; Williams, 1977).
Hegemony is a provisional alliance of social ideas
which exerts social authority over other subordinate
groups. The dominant win and shape consent so their
ideas appear permanent and natural, beyond particular interests (Hall, 1977). Hegemony has to be
won, reproduced and sustained. It is a 'moving
equilibrium', containing elements favourable or
unfavourable to this or that tendency (see Hall, et
al, 1976). It is unusual then rather than usual for
a particular view to dominate for long. Such
periods may be illustrated by a consistency of
theory and practice - the integration of policies
which we have exemplified in this chapter. What we
have also exemplifed is the breakdown that consistency - through redefinition, additional demands or
a radical reappraisal of where society should go.
These reappraisals in themselves point clearly to
particular views of need and justice, the clarity of
which becomes more opaque with time. Thus the new
Poor Law recognised only the needs of the deserving
poor. Rights to justice were limited to those
enjoyed under the law. Differential citizenship
ensured that those rights were not universal. All
other 'needs' were wants and as such immoral and
unrecognised. The social engineers of the post-1945
period saw that each citizen (now a universal cate-

gory) had basic needs which s/he must satisfy alone
or with state assistance. These social rights were
based on fairness but that co-existed with the
liberal capitalist notion of desert in that those
who succeeded in obtaining over and above their
necessary requirements were entitled to keep it. In
the last 10 years, although it is possible to say
the some of the morality of the new Poor Law has
been reintroduced, especially and most cruelly with
the poorest in society, it is rather a reversal of
the emphasis of the social engineers which domi-
nates. Desert is established as at least a strong a
principle as basic need fulfilment. In line with
this emphasis, equality of opportunity rather than
fairness is the dominant view of justice. Further,
just as the minimum benefit provided by the state is
seen as a safety net so to is equality under the
law. Two points follow. The benefit that constitutes the safety net is being eroded under the
principles of the new morality. And the law is
increasingly used in a partial way. These points
take us, however, into a broader discussion which
lies beyond the scope of this book. But under any
system, morality and law are not neutral devices but
are used to further particular purposes and ends.
Resource allocation is a moral as well as social and
technical issue.

Chapter Six

THE GEOGRAPHY OF WELFARE

Joint Care and Local Initiatives

We opened this book by suggesting that we took a
specific view of the meanings of both 'geography'
and 'national health'. We accept that these distinctive starting-points have given our work a
particular, and what would appear in places to be a
non-geographical, emphasis. We shall deal with what
this means for the geographical shortly but we wish
to start this final chapter by briefly extending
some of the arguments put forward in Chapter Five on
integrated approach to health and welfare.

We argued that most, if not all, policy initiatives in the health and welfare fields are based at
root on some general integrating principles. We
illustrated this argument by examining moral-
economic integration and the new Poor Law, socially
engineering integration and welfare state planning
and the financial-moral integration of the health
and welfare reductions in the last 10 years. In the
first two cases, we noted how the principles of
integration became gradually diluted as practical
problems of provision intruded into the more theore-
tical considerations. That dilution was not noted
with the last example as its course has perhaps not
yet been run. We did note, however, how the unintended consequences of particular actions tended to
reverberate around the entire public expenditure
system. So a cut in one department's budget may in
effect increase expenditure in another, particularly
if this latter agency has a legal or statutory duty
to provide services at a particular level or time.
Reductions, in hospital expenditure are associated
with increases in spending on family practitioner
services, while the removal of industrial aid and
subsidies may be linked, though increased redundancy

206

and unemployment, to additional housing subsidies in the forms of rent and rate rebates. Further, the cutbacks in the housing construction and improvement budgets has led, especially in certain places, to greater expenditure and strain on provision for social service departments. One solution to 'solve' these unintended difficulties is to attempt to cash-limit more and more services. As we noted in Chapter Five, the voucher scheme for primary care, floated by the Conservative Government, is one such attempt. Another is to try and make individuals, including the disadvantaged, more self-reliant. Welfare payments become increasingly parsimonious to this end. A further measure is to try to rationalise the use of resources across the service boundaries of different agencies. In this and within the context of a general cost-containment policy, we witness a form of practical integration. It is known as joint care planning (see also Chapter 5).

In fact, the desire for joint planning emanated from the more optimistic, social engineering period of the early 1970s. A working party was then established to examine the possibilities of collaboration between the NHS and local government departments (DHSS, 1973; George, 1985). Such collaboration would avoid duplication and help establish comprehensive and coordinated services. Two specific measures were introduced, the first being the creation of joint consultative committees (JCCs). These were meant to cover education, housing and environmental health as well as health care and personal social services but the former had not really developed. It is suggested that there is less overlap and less common interests (Walker, 1984) but as we have shown in Chapters Three and Four such is not the case. Further, other central government departments would be involved, for example Department of the Environment, Department of Education and Science. Such liaison has perhaps produced visions of a bureaucratic nightmare, this being sufficient to declare a difficult job an impossible one, ensuring no progress towards service integration. The need for such integration was also argued for by the Royal Commission on the NHS (Great Britain, 1979). Secondly, joint finance is available. This applies to the health-personal social services link and its expenditure is guided by joint care planning teams (JCPTs).

Joint finance means that a proportion of the health budget (around 1 per cent) is set aside each year for projects in the personal social services.

It is a way of encouraging not only coordination but
a commitment to a cheaper kind of care, that in the
community. Such care should not of course be des-
pised but its use as a cost saving strategy must be
recognised. It is thus a practical way of switching
care from institutions to community-based services
(Sargent, 1979; Booth, 1981; Walker, 1984). But
although the uptake of available funds from joint
finance has been large, joint planning has not been
very successful as yet. These are technical and
organisational difficulties in, for example, the
lack of co-terminous boundaries, accounting and
constitutional differences, and the small scale and
short duration of funding (see Glennerster et al,
1983). There are philosophical difficulties in that
planning cannot overcome differences in service
structure and professional outlooks, inadequacy of
service coverage and the lack of accountability to
clients and community (Walker, 1984). There have
also been practical difficulties in that local
authorities have tended to use joint finance for
schemes already set up but under threat from rate
capping. These fears have also meant that capital
rather than revenue (staff) projects have been
preferred (Social Services Committee, 1985; George,
1985). George also points out that there is a
conflict of aims between the health and local authorities which, it may be argued, stems very much from
their organisational natures. The local authorities
thus emphasise domiciliary and day care services
while the institutional bias of the NHS and its
desire to deinstitutionalise its priority groups
have led to its wish to expand residential accommodation. These differences are likely to remain
unresolved especially in cash-limited times in which
what is already done and for which a statutory
responsibility is held will be pursued at the
expense of any attempt to redefine who does what and
how. This redefinition is particularly difficult to
achieve anyway. Even where a structure has been
evolved to develop integrated and coordinated welfare and health systems as in Northern Ireland,
progress in setting up 'programmes of care' has been
limited. Coordination has been improved but changes
in professional attitudes and practice take a long
time to occur (see Birrell and Williamson, 1983).
Further, what may be clear at the centre often
becomes murky at the periphery and while the need
for obtaining and ways to implement integrated
practice may be fully understood at the centre, at
the local level - where care is given - things may

be more problematic.

It is of course at the local level that all policy must be implemented and where its effects are surely felt. While in the main national policy has been towards greater individual provisions and self-reliance in health and welfare, the effects have not been uniform. The effects of curtailing state provision are more likely to be felt in places and among groups that do not have the resources to become self-reliant. Class and industrial structure will, therefore, in large measure determine the local effects of centrally imposed cuts. It may even be argued that the relatively poor in affluent areas suffer less than their compatriots with simi-lar attributes in poorer areas. If the relatively wealthy take out private insurance and are treated by the resources of commercial medicine then there may be more NHS facilities available for the rela-tively poor. We should note, however, that the relatively wealthy do not stop using NHS facilities. They merely use the private sector for particular treatments. Further, there have been suggestions that private facilities should be incorporated into RAWP calculations for allocating funds for public facilities. If that happens areas in the outer south-east will lose NHS resources.

But class and industrial structure do not entirely or directly determine the effects of public expenditure reductions. The picture is complicated by the local state which may set up initiatives which attempt to militate locally against the cuts. Such initiatives are limited by the resources that can be generated locally and by central government rate-capping. They are also limited by the obvious fact that so much health care provision is beyond local control and influence. Some local authorities have concentrated on attempting to maintain their housing programmes, as in Liverpool, an important strategy given the relationship between health and housing. Others have tried to improve mobility through the subsidising of bus fares as in South Yorkshire. In part of South Yorkshire, in Sheffield, attempts have been made to improve hous-ing conditions by bringing together tenants and direct labour workers and to initiate client-sensi-tive welfare provision (see Alcock & Lee, 1981; Deacon, 1983). Several London boroughs - Hackney & Islington to the fore - have followed the example of Walsall in decentralising various services and running them from neighbourhood offices. These are generic in the sense of dealing with housing, wel-

fare and educational problems. Their success relies
on a change in the attitudes of professionals providing advice and services. The Walsall scheme
foundered on recruiting sympathetic personnel even
before Labour lost its electoral majority (see
Seabrook, 1983). Hackney's attempt at decentralising services and council power (excluding the raising of rates) to neighbourhood centres with committees made up of councillors, tenants' and residents'
association members and community representatives
was challenged by the boroughs employees whose
organisations saw the issue as one that might weaken
their role and the conditions of service of the
workforce. The localisation of service provision
through decentralisation seems, therefore, to be set
against the sectional interests of workers in the
welfare services. This problem seems to be a micro-
cosm of the difficulties facing socialist provision
and principles of allocation. Can a socialist
society derive from a transformation of the means of
provision or must it simply come from the transformation of work relationships?

Socialist Integration and Planning

In Chapter Two, we set out bases of a socialist
system of allocation, with needs being seen in
relationship to the avoidance of harm, and with the
system being necessarily interventionist and formulated with respect to the goals of a socialist
society. These goals, unlike Labour's alternative
economic strategy, require the integration of social
and economic policy and within that broad aim the
integration of the welfare services. The broad
policy integration enables an answer to be given to
the question which ended the last section. It is
necessary to transform the provision of welfare as
well as work relationships. But the process is
extremely complicated. As Ferge (1979) shows in the
Hungarian context, a change in ownership does not
necessarily result in socialist relations. The
former capitalist pattern continues to exist under
the 'socialist' conditions. We do not wish to
become embroiled in this debate (see Corrigan et al,
1978; Deacon, 1983). Rather we concentrate on
socialist integration in health and welfare.
     Even in this we must recognise that broad issues
concerning the nature of humankind and society are
involved. "Societal or structural policy is a
fairly ambitious concept. It implies the project of
deliberately changing the profile of a society, of

altering basic social, human relations" (Ferge, 1979, 55). As we argued in Chapters One and Two, social policy is based on the rights and obligations of people to be cared for by others and to care for others on the basis of need. We obtain these rights and obligations as members of collectivities (usually citizens of states) with these rights and obligations being manifested at the level of the individual. This means that socialist integration concerns questions of 'the individual', 'citizen', 'nation' and so on. These terms have been given particular bourgeois meanings but such meanings should not become universal ones by default. They must be given socialist meanings because they form a basis for the most fundamental of integrations, that of individuals to one another in a collectivity. In this view, we support, therefore, the rather out-moded socialist humanism of Goldman which emphasises values which help people make sense of the world and their place in it. Goldman (1966, 40) argues for

> an integration of the major values inheri-
> ted from middle-class humanism (universa-
> lity and individual freedom, equality, the
> dignity of the human person, freedom of
> expression) so as to endow them, for the
> first time in the history of humanity, with
> a quality of authenticity, instead of the
> purely formal status that they had pre-
> viously been granted in capitalist society.

These values may be seen as bases for caring and Goldman (1966, 50) goes on:

> Socialist society was expected to restore
> and further develop the values of Western
> Humanism, since it would not only strip
> them of their merely formal character by
> suppressing all exploitation and class
> distinctions, but also bind them organi-
> cally to a community both truly human and
> fully conscious of those trans-individual
> values which would be liberated at last
> from the heavy handicaps that poverty and
> exploitation had imposed in the pre-capi-
> talist periods of history.

But how might such values - bases of caring - be inculcated? Collective activities may militate against individualism and consumerism (see Walker, 1984), but how in turn are such activities encouraged? What, for example, is the role of the state in formulating socialist relations of welfare and in enhancing socialist integration? We have already said that socialist practice is interventionist and based on planning but these activities must not be foisted on an unwilling and passive population. Under capitalism, the state may in the field of health further people's dependence, commodify health care and enhance the conditions to ensure accumulation and exploitation (Illich, 1977; Renaud, 1975; Navarro, 1974) but it need not be so. Socialist planning must be popular in the sense of being based on people's experiences and being democratic and decentralised. The need for populism has at least two implications for socialist integration. First, it must recognise that policy must be formulated in a situation where there are divisions and prejudices, arising from experiences. It is unlikely that mere exhortation will make these unpalatable matters go away. Secondly, it is unlikely that the cause of democratic planning will be greatly furthered by existing institutions or organisations. Neighbourhood centres were seen as one possible way of instituting such planning. We should not decry the locality as a potential arena for plan-making and integration (both for service provision and for commitment to caring). The neighbourhood is where most social interactions occur and where most needs are met (if not serviced). It also has great potential as a basis for protest and struggle against existing institutions (see Habermas, 1979; Eyles & Evans, 1986). And if existing institutions are unlikely to enhance socialist integration, then struggle is likely to be extremely important.

If socialist planning is to achieve an equitable, just society based on needs, how might that be achieved within the general purview of integrated social and economic policy? First, we must remember that its achievements should be monitorable and measureable. This requires not only a clarity of aims but also techniques to demonstrate the differences between particular time-periods. These techniques include quality of life measures of the social indicator type which would inform on whether group A or place X was better or worse off than before. But as Carlisle (1972) notes, indicators can do more than this. By linking them to a model of society, they can measure progress towards or away from particular societal goals - equality of outcome, social utility of production and so on. Indeed, such indicators would be the basis of a system of social accounting to allow for the allocation of resources on social rather than market-based criteria (see also Archibugi, 1978; Walker, 1984).

Secondly, under a socialist system, resources are simply ordered differently: they are not infinite so it must be decided how much priority is to be given to health and welfare provision.

This question of priority is only one of those which Deacon (1983) suggests needs to be asked about socialist social policy. His six questions are

1. What priority would be afforded to social welfare provision within a socialist society and what resources would be available to satisfy the priority?

2. What form of control over institutions of welfare might be established within a socialist society?

3. What would be the balance between the state, the market, the workplace, the community and the family in the provision of social welfare?

4. What might become of the existing relationships between users, providers and administrators of the welfare services?

5. What system of distribution and rationing of services might be developed?

6. What changes will have taken place in the nature of family life and in the sexual divisions of labour and what impact would this (sic) have had on the forms of welfare policy and provision? (Deacon, 1983, 14).

Thus from these questions we may see that the socialist integration of welfare provision requires not simply the conjoining of health care, social services, housing and so on but also establishing the interrelationships between providers and clients/patients and between different forms of care and determining the nature of control and distribution. Just because the inequities of the present capitalist system are recognised does not mean that hard questions on democratic v. centralised control, on universal v. special access and on institutional v. community care go away. They remain to be answered, albeit from a different viewpoint and on the basis of a different set of principles. And there also remains the general resource constraint.

Welfare Geography and the Geographies of Welfare

We wish now to turn to issues of a more academically geographical nature and return to the sub-theme of the book, welfare geography. We stated that the

purpose of the book was to examine the shape and
texture of the geography of the national health.
'Geography' was taken as it is traditionally to mean
the patterning of conditions across territories and
as a component of policy (as many policy initiatives
in the health and welfare fields have spatial
bases). But while a geographical perspective was
seen as a necessary frame of reference, it must be
part and parcel of a broad-based examination of
health. For this reason, the geography of the
national health may appear to be underplayed in
certain parts of this book. Where policy has
attempted to bring about a state of affairs that
obtains in the whole country, we have felt it necessary to describe and understand those policies,
whatever happens to the 'geography'. But as we hope
will be seen in earlier chapters, these universalising policies have geographically as well as socially
uneven effects. Because of the relative wealth of
localities, the new Poor Law meant better or worse
poverty relief and health care provision depending
on location. The norms of the Hospital Plan of the
1960s did not result in equality of bed provision in
different places because of organisational problems,
uneven progress and financial stringency. It was,
however, in our third example - the period from the
mid-1970s to the present - that the geography of the
national health stands out. Policy in this period
is not guided by universalist, integrating principles as those which tried to equalise provision but
by cost containment. This containment, along with
the encouragement of commercial medicine, has meant,
except at the margins, that geographical differences, based on past provision, class and industrial
structure, have remained largely untouched while the
actual quantity and quality of provision in nearly
all places have worsened.
    In taking this policy view of the geographical -
in terms of administrative entities - we have departed somewhat from the approach traditionally adopted
in welfare geography. To be sure, administrative
units are used but they are the containers of attributes which themselves vary for better or worse
across those units. While it was certainly not the
stated intention of the leading geographer in this
area of interest, Smith (1977), his own work and
most of that of other welfare geographers with the
possible exception of Kirby (1982) has concentrated
on describing states of inequality and disadvantage
in various places. In fact, given the nature of
Smith's book <u>Human geography: a welfare approach</u>

with its concentration on welfare and marxian economics and their relation to location, his project
stands or falls on how well his chosen theories
explain the phenomena he describes. We do not wish
to become embroiled in an argument that peaked in
the review of the work some years ago, although in
our view the connections between explanation and
descriptions are not always clear.

We wish rather to return to the stated intention
- indeed what we could call the unfulfilled promise
- of what Smith's refers to as the welfare theme or
approach. We have already noted the emphasis on
description at expense of other identified basic
tasks, namely evaluation and prescription. In other
words, there is less attention paid to the policy
process and mechanisms by which resources are and
may be allocated to particular groups and territories, although general remarks about different social
systems and their inequalities are found in later
work (Smith, 1979). This failure is unfortunate
given the quotations from welfare economics, in
which the word geography is substituted for economics, at outset of the section of 'the welfare
theme'. Thus "welfare geography is that part of
geography where we study the possible effects of
various geographical policies on the welfare of
society" (Smith, 1977, 6). But policy becomes lost
possibly in Smith's desire to recognise the primary
concern of human geography. But it is a rather
traditional primary concern that becomes emphasised,
namely areal differentiation, couched to be sure in
terms of life chances, but areal differentation
nonetheless.

Saying this does not mean that we deride the
work of Smith or of the welfare approach. In recognising the significance of welfare, in using social
indicators and in seeing that geography has much to
be with morality as with scientific enquiry, Smith's
work is of enduring importance. But in following
Lasswell's (1958) view that politics is the study of
who gets what, when and how and Samuelson's (1973)
that economics is about what, how and for whom in
describing human geography as who gets what, where
and how (Smith, 1974), he changes the emphasis. In
remaining true to a traditional conception of the
subject, he abrogates concern for the policy arena.
And this arena is surely the correct focus for any
welfare theme. More clumsily but accurately put,
the welfare approach in geography stresses where is
what which whom gets. The crucial omission is
'how'. In other words, what is missing are the

decisions and mechanisms which influence who gets
what, where. And as we tried to show in Chapter
Two, policy is simply the visible outcome of the
operation of allocative mechanisms that are themselves shaped by power relations and conceptions of
need and justice. Further as we illustrated in
Chapters Two and Five and earlier in this Chapter, a
consideration of 'how' of policy does not have to
begin and end with what already exists, with what is
already formulated. By a process of evaluation,
alternative states and policies may be envisaged and
expounded.

    Our examination of the geography of the national
health is, therefore, also an argument for the
reconstruction of welfare geography. We believe
that its emphases on a broadly defined well-being
(integrated lives which can be used as a basis for
examining how integrated policies are), comparative
analysis (not much in evidence in this book) and
social indicators and monitoring are well worth
retaining. So too is the question: who gets what,
where and how. But a geography which is part of a
broader social science interest in policy, evaluation and prescription must place different emphases
on the various elements in the question from those
of the welfare approach. These different emphases
are illustrated by the different chapters of this
book. Because we are geographers, 'where' is the
underlying theme, colouring much else. Its explicit
treatment may be found particularly in Chapters Two,
Three, Four and Five. By 'who', we mean individuals
and groups (in space) and their attributes. They
may be the recipients of policy initiatives. Or
they may be ignored as such. In Chapters Three and
Four, we examined 'who' in the context of those ill
and poor people in Britain in need of health care.
By what, we mean the resources which go to make life
satisfactory or unstafisctory. They are more than
state aid and as we saw in Chapter One much depends
on how we define health, care and resources.
Indeed, the importance of such definitions makes the
investigation of geographies of welfare of overriding importance because these definitions will
vary from society to society complicating but potentially enriching the process of evaluation. And
studies of geographies of welfare are also required
by considering the last element in the question,
'how'. Different policy options, allocative mechanisms, power relations and conceptions of need and
justice are found in different societies. We have
only examined one. But how do these different parts

216

of the how question come together and what do they
mean for present and future health and welfare
provision and distribution? Some excellent comparative analysis of welfare (eg Titmuss, 1973;
Higgins, 1981) point the way but one challenge
remains the use of such analyses - and not just in
terms of per capita expenditure, service provision
ratios - to illuminate geographies of welfare.

BIBLIOGRAPHY

Abel-Smith, B (1964) The hospitals 1800-1948,
    Heinemann
Abel-Smith, B (1970) Public expenditure on the
    social services, Social Trends, 1, 12-20
Abel-Smith, B & R M Titmuss (1956) The cost of the
    national health service in England and Wales,
    Cambridge UP
Abrams, M (1973) Subjective social indicators,
    Social Trends, 4, 35-50
Abrams, P (1977) Community care, Policy and Politics
    6
Acheson, D (1981) Primary health care in inner
    London, DHSS
Acton, H B (1971) The moral of markets, Longman
Addison, P (1975) The road to 1945: British poli-
    tics and the second world war, Quartet
Airth, A D & D J Newell (1962) The demand for hospi-
    tal beds, University of Durham
Alcock, P & P Lee (1981) The socialist republic of
    South Yorkshire, Critical Social Policy 1(2),
    72-93
Alford, R (1975) Health care politics, University of
    Chicago Press
Allan, G (1983) Informal networks of care, British
    Journal of Social Work 13, 417-33
Allen, T (1973) A tale of high living, The Guardian
    11 January
Allsop, J (1984) Health policy and the NHS, Longman
Archibugi, F (1978) Capitalist planning in question,
    in S Holland (ed) Beyond capitalist planning,
    Basil Blackwell

Bacon, R & W Eltis (1976) Britain's economic prob-
    lems, Macmillan
Barnes, J (1975) Educational priority, vol 3, HMSO
Barr, A (1957) The population served by a hospital

group, Lancet 2, 1105

Barry, B (1973) The liberal theory of justice, Clarendon Press

Bayley, M (1973) Mental handicap and community care, RKP

Beeson, P B (1980) Changes in medical therapy during the past half century, Medicine 59, 79-89

Bell, C (1968) Middle class families, RKP

Benn, S I & R S Peters (1959) Social principles and the democratic state, Allen & Unwin

Berger, P B, Berger & H Kellner (1974) The homeless mind, Penguin

Berthoud, R & J C Brown (1981) Poverty and the development of anti-poverty policy in the UK, Heinemann

Beveridge, W H (1943) The pillars of security, Macmillan

Birrell, D & A Williamson (1983) Northern Ireland's integrated health and personal social services structure, in A Williamson and G Room (eds) Health and welfare states of Britain, Heinemann

Blaug, M (1964) The poor law re-examined, Journal of Economic History 24

Blaxter, M (1976) Social class and health inequalities, in C Carter and J Peel (eds) Equalities and inequalities in health, Academic Press

Blaxter, M & E Paterson (1982) Mothers and daughters, Heinemann

Booth, T A (1981) Collaboration between the health and social services, Policy and Politics 9, 23-49

Bradley, J E, A M Kirby & P J Taylor (1978) Distance decay and dental decay, Regional Studies 12, 529-40

Bradshaw, J (1972) The concept of social need, New Society 30 March, 640-3

Braybrooke, D (1968) Let needs diminish that preferences may prosper, in N Rescher (ed) Studies in moral philosophy, Basil Blackwell

Brearley, P (1978) The social context of health care, Blackwell

Brenner, M H (1979) Mortality and the national economy, Lancet ii, 568-73

British Medical Journal (1978) Two for the price of one, British Medical Journal 2, 113

Brookfield, H C (1975) Interdependent development, Methuen

Brotherston, J H F (1971) Change and the national health service, in A Gatherer and M D Warren (eds) Management and the health services, Pergamon

Brown, C (1984) Black and white Britain, Heinemann

Bibliography

Brown, R G H (1979) Reorganising the health service,
    Martin Robertson
Bruce, M (1971) The coming of the welfare state,
    Batsford
Brundage, A (1972) The landed interest and the new
    poor law, English Historical Review 87, 27-48
Butler, J (1973) Family doctors and public policy,
    RKP
Butler, J R & M S B Vaile (1984) Health and health
    services, RKP
Butler, S M (1982) Enterprise zones, Heinemann
Butts, M, D Irving & C Whitt (1981) From principles
    to practice, NPHT

Campbell, A V (1978) Medicine, health and justice,
    Churchill-Livingstone
Campbell, T D (1974) Humanity before justice,
    British Journal of Political Science 4
Carlisle, E (1972) The conceptual structure of
    social indicators, in A Shonfield & S Shaw (eds)
    Social indicators and social policy, Heinemann
Carstairs, V (1981) Multiple deprivation and health
    state, Community Medicine 3, 4-13
Castle, B (1980) The Castle diaries, 1974-76,
    Weidenfeld & Nicolson
Central Health Services Council (1969) The functions
    of the district general hospital, HMSO
Central Statistical Office (CSO) (1975) Social
    Trends, HMSO
Central Statistical Office (CSO) (1985) Annual
    abstract of statistics, HMSO
Charlton, J et al (1983) Geographical variation in
    mortality from conditions amenable to medical
    intervention in England and Wales, Lancet ii,
    691-6
Chilvers, C (1978) Regional mortality 1969-73,
    Population Trends, 11, 16-20
Cleland, E A, R J Stimson & A J Goldworthy (1977)
    Suburban health care behaviour in Adelaide,
    Flinders University, Centre for Applied Social
    and Survey Research, Monograph Series 2
Cobb, S (1976) Social support as a moderator of life
    stress, Psychosomatic Medicine 38, 300-14
Cochrane, A L (1972) Effectiveness and efficiency,
    NPHT
Conway, J & P Kemp (1985) Bed and breakfast, SHAC
Conyers, D (1982) An introduction to social planning
    in the third world, Wiley
Cook, P J & R O Walker (1967) The geographic distri-
    bution of dental care in the UK, British Dental
    Journal 122, 441-7, 494-9, 551-8

Bibliography

Cooper, M H (1975) Rationing health care, Croom Helm
Cornwell, J (1984) Hard-earned lives, Tavistock
Corrigan, P (1977) The welfare state as an arena of
    class struggle, Marxism Today 21(3)
Corrigan, P, H Ramsay, & D Sayer (1978) Socialist
    construction and marxist theory, Macmillan
Craig, J & A Driver (1972) The identification and
    comparison of small areas of adverse social
    conditions, Applied Statistics 21, 25-35
Crawford, R (1980) Healthism and the medicalisation
    of everyday life, International Journal of Health
    Services 10, 365-88
Crosland, C A R (1956) The future of socialism, Cape
Crosland, C A R (1974) Socialism now, Cape
Cross, M (1982) For sale New Scientist 94, 1309,
    710-13
Cullinan, T & J Treuherz (1981) Ill in east London
    1979-81, Department of Environmental and Preven-
    tive Medicine, St Bartholomew's Hospital Medical
    College
Cullingworth, J B (1973) Problems of an urban
    society, Allen & Unwin
Culyer, A J (1976) Needs and the NHS, Martin
    Robertson
Curtis, S (1983) Intra-urban variations in health
    and health care, Department of Geography, Queen
    Mary College
Curtis, S & K J Woods (1984) Health care in London,
    in M Clarke (ed) Planning and analysis in health
    care systems, Pion

Dahl, R A & C E Lindblom (1963) Politics, economics
    and welfare, Harper
Daniels, N (1975) Reading Rawls, Basil Blackwell
Davies, B (1968) Social needs and resources in local
    services, Michael Joseph
Davies, B (1977) Needs and outputs, in H Heisler
    (ed) Foundations of social administration,
    Macmillan
Day, P & R Klein (1985) Towards a new health care
    system? British Medical Journal 291, 1291-3
Deacon, B (1983) Social policy and socialism, Pluto
    Press
Department of Health and Social Security (1973) A
    report from the working party on collaboration
    between the NHS and local government, HMSO
Department of Health and Social Security (1976a)
    Guide to planning in the NHS, DHSS
Department of Health and Social Security (1976b)
    Priorities for health and personal social
    services in England, HMSO

Bibliography

Department of Health and Social Security (1976c)
    Prevention and health, HMSO
Department of Health and Social Security (1977a)
    Health and personal social services statistics,
    1976, HMSO
Department of Health and Social Security (1977b) The
    way forward, HMSO
Department of Health and Social Security (1979)
    Patients first, HMSO
Department of Health and Social Security (1980)
    Inequalities in health, HMSO
Department of Health and Social Security (1981) Care
    in action, HMSO
Department of Health and Social Security (1984) The
    health service in England, annual report, HMSO
Department of the Environment (1977a) Change or
    decay, HMSO
Department of the Environment (1977b) Policy for the
    inner cities, HMSO
Digby, R (1982) The poor law in nineteenth-century
    England and Wales, Historical Association,
    General Series 104
Dijksterhuis, E J (1961) The mechanisation of the
    world picture, Oxford UP
Dingwall, R (1976) Aspects of illness, Martin
    Robertson
Dixon, B (1978) Beyond the magic bullet, Allen &
    Unwin
Dollery, C T (1978) The end of the age of optimism,
    NPHT
Donaldson, L J (1976) Urban and suburban differen-
    tials, in C Carter & J Peel (eds) Equality and
    inequality in health, Academic press
Donnison, D (1975) Equality, New Society 20
    November, 422-4
Donnison, D & P Sotto (1980) The good city,
    Heinemann
Donovan, J (1984) Ethnicity and health, Social
    Science and Medicine 19, 653-70
Donovan, J (1985) Black people's health, unpublished
    PhD thesis, University of London
Downie, R S & E Telfer (1980) Caring and curing,
    Methuen
Doyal, L (1979) The political economy of health,
    Pluto Press
Dubos, R (1959) Mirage of health, Harper & Row
Dunkley, P (1973) The landed interest and the new
    poor law: a critical note, English Historical
    Review 88, 836-41

Eckstein, H (1958) The English health service,

Harvard UP

Edwards, J & R Batley (1978) The politics of positive discrimination, Tavistock

Ehrenreich, B & J Ehrenreich (1978) Medicine and
social control, in J Ehrenreich (ed) The cultural
crisis of modern medicine, Monthly Review Press

Elwood, P C & J E J Gallacher (1984) Lead in petrol
and levels of lead in the blood, Journal of
Epidemiology and Community Medicine 38, 315-8

Emmet, D (1966) Rules, roles and relations,
Macmillan

Eyles, J (1979) Area-based policies for the inner
city, in D T Herbert and D M Smith op. cit.

Eyles, J (1985a) Senses of place, Silverbrook Press

Eyles, J (1985b) From equalisation to rationalisation, Australian Geographical Studies 23, 243-69

Eyles, J (1986) Poverty, deprivation and social
planning, in M Pacione (ed) Progress in social
geography, Croom Helm

Eyles, J & J Donovan (1986) Regional variations in
perceptions and experiences of health and health
care, ESRC End of Research Report

Eyles, J and M Evans (1987) Popular consciousness,
moral ideology and locality, Society and Space 5

Eyles, J & K J Woods (1983) The social geography of
medicine and health, Croom Helm

Eyles, J & K J Woods (1986) Who cares what care: an
inverse interest law? Social Science and
Medicine 21

Farrow, S (1983) Monitoring the health effects of
unemployment, Journal of the Royal College of
Physicians 17, 99-105

Ferge, Z (1979) A society in the making, Penguin

Fieghen, G C, P S Lansley & A D Smith (1977) Poverty
and progress in Britain 1953-73, Cambridge UP

Fishman, W (1979) The streets of east London,
Duckworth

Flinn, M (1976) Medical services under the new poor
law, in D Fraser (ed) The new poor law in the
nineteenth century. Macmillan

Forder, A (1984) Theories of welfare, RKP

Forster, D P (1977) Mortality,.morbidity and
resource allocation, Lancet 8019, 997-8

Forsyth, G & R F L Logan (1960) The demand for
medical care, Oxford UP

Fox, A J & P O Goldblatt (1982) Socio-demographic
mortality differentials: longitudinal study
1971-5, HMSO

Freidson, E (1970) Profession of medicine, Dodd Mead

Friedmann, M (1962) Capitalism and freedom,

Bibliography

University of Chicago Press

Galbraith, J K (1958) The affluent society, Hamish
    Hamilton
Galbraith, J K (1967) The new industrial state,
    Deutsch
Galbraith, J K (1974) Economics and public purpose,
    Deutsch
Gamble, A (1980) The free economy and the strong
    state, Socialist Register 16, 1-25
Gamble, A & P Walton (1976) Capitalism in crisis,
    Macmillan
Garfinkel, H (1967) Studies in ethnomethodology,
    Prentice-Hall
Geary, K (1977) Technical deficiencies in RAWP,
    British Medical Journal 21 May
George, C (1985) The redefinition of community care,
    Papers and Proceedings of the First IBG-AAG
    Symposium on Medical Geography, Univ of
    Nottingham
George, V & P Wilding (1976) Ideology and social
    welfare, RKP
Giggs, J (1979) Human health problems in urban
    areas, in Herbert & Smith, op. cit.
Gittus, E (1965) An experiment in the definition of
    urban sub-areas, Transactions of the Bartlett
    Society 2, 109-35
Gittus, E (1976) Deprived areas and social. planning,
    in D T Herbert & R J Johnston (eds) Social areas
    in cities vol 2, Wiley
Glennester, H, N Korman & F Marslen-Wilson (1983)
    Planning for the priority groups, Martin
    Robertson
Gough, I (1979) The political economy of the welfare
    state, Macmillan
Gould, D (1971) A groundling's notebook, New
    Scientist 51, 217
Gravelle, H S E, G Hutchinson, J Stern (1981)
    Mortality and unemployment, Lancet iii, 675-9
Gray, J A M (1979) Man against disease, Oxford UP
Graycar, A (1983) Informal, voluntary and statutory
    services, British Journal of Social Work 13,
    379-93
Great Britain (1942) Social insurance and allied
    services, (Beveridge Report) cmnd 6404. HMSO
Great Britain (1944) A national health service, HMSO
Great Britain (1956) Committee of enquiry into the
    cost of the national health service (Guillebaud
    Report) HMSO
Great Britain (1975) Better services for the
    mentally ill, HMSO

Bibliography

Great Britain (1976) <u>Sharing resources for health in England</u>, HMSO
Great Britain (1979) <u>Royal commission on the NHS</u>, HMSO
Greater London Council (GLC) (1984) <u>London's docklands</u>, GLC
Gregg, P (1982) <u>A social and economic history of Britain 1760-1980</u>. Harrap
Grey, A M H & A Topping (1945) <u>The hospital survey</u>. HMSO
Griffith, B & G Rayner (1985) <u>Commercial medicine in London</u>, GLC

Habermas, J (1976) <u>Legitimation crisis</u>, Heinemann
Habermas, J (1979) <u>Conservatism and capitalist crisis</u>, New Left Review 115
Hall, D (1983) <u>The cuts machine</u>, Pluto Press
Hall, J (1976) Subjective measures of quality of life in Britain, 1971-5, <u>Social Trends</u> 7, 47-60
Hall, P (1974) <u>Urban and regional planning</u>, Penguin
Hall, P (1981) <u>The inner city in context</u>, Heinemann
Hall, R & P Ogden (1983) <u>Europe's population in the 1970s and 1980s</u>, QMC
Hall, S (1977) Culture, the media and the 'ideological effect', in J Curran (ed) <u>Mass communication and society</u>, Arnold
Hall, S, J Clarke & B Roberts (1976) Subculture, culture and class, in S Hall et al <u>Resistance through ritual</u>, Hutchinson
Halsey, A (1972) <u>Trends in British society since 1900</u>, Macmillan
Ham, C (1981) <u>Policy-making in the NHS</u>, Macmillan
Hamnett, C (1979) Area based explanations, in Herbert & Smith <u>op. cit.</u>
Hampshire, S (1972) Review of 'a theory of justice', <u>New York Review of Books</u>, 24 February
Hannock, C P (1973) <u>Fit and proper persons</u>, Arnold
Hannock, R (1971) Marx's theory of justice, <u>Social Theory and Practice</u> 1(3), 65-71
Hart, J T (1971) The inverse care law, <u>Lancet</u> 1, 405-12
Harvey, D (1973) <u>Social justice and the city</u>, Arnold
Hayek, F (1944) <u>The road to serfdom</u>, RKP
Hayek, F (1960) <u>The constitution of liberty</u>, RKP
Hayek, F (1966) The principles of liberal social order, <u>Il Politico</u> 31, 601-17
Hayek, F (1968) <u>The confusion of language in political thought</u>, Institute of Economic Affairs
Hayek, F (1976) <u>Law, legislation and liberty vol 1: the mirage of social justice</u>, RKP
Haywood, S & A Alaszewski (1980) <u>Crisis in the</u>

health service, Croom Helm

Heller, A (1976) The theory of need in Marx, Allison & Busby

Henriques, U R Q (1979) Before the welfare state, Longman

Herbert, D T & D M Smith (eds) Social problems and the city, Oxford UP

H M Treasury (1979) Government expenditure plans 1979-80 to 1982-3, HMSO

H M Treasury (1985) Government expenditure plans 1985-6 to 1987-8, HMSO

Herrington, J (1984) The outer city, Harper & Row

Herzlich, C (1973) Health and illness, Academic Press

Higgins, J (1981) State of welfare, Basil Blackwell and Martin Robertson

Hill, J D, J R Hampton & J R A Mitchell (1978) A randomised trial of home-versus-hospital management for patients with suspected myocardial infarction, Lancet 1, 837-41

Hodgkinson, R (1967) The origins of the national health service, Wellcome Historical Medical Library

Hoggart, R (1957) The uses of literacy, Penguin

Holterman, S (1975) Areas of urban deprivation in Great Britain, Social Trends 6, 33-47

Honoré, A M (1968) Social justice, in R S Summers (ed) Essays in legal philosophy, Clarendon Press

Horwitz, A (1978) Family, kin and friend networks in psychiatric help-seeking, Social Science and Medicine 12, 297-304

Howe, G M (1970) A national atlas of disease mortality in the United Kingdom, RGS

Howe, G M (1972) Man, environment and disease in Britain, Penguin

Hunt, K (1985) Objective and subjective measures of social well-being in Portsmouth, Queen Mary College, Department of Geography and Earth Science, Undergraduate Dissertation

Hunt, S & J McEwen (1980) The development of a subjective health indicator, Sociology of health and illness 2, 231-46

Hunt, S et al (1981) The Nottingham health profile, Social Science and Medicine 15A, 221-9

Hunter, D (1979) Coping with uncertainty, Sociology of Health and illness 1

Hutchinson, T W (1970) Half a century of Hobarts, Institute of Economic Affairs

Iliffe, S (1983) The NHS, Lawrence & Whishart

Illich, I (1975) Medical nemesis, Marion Boyars

Illich, I (1977) Limits to medicine, Penguin
Illsley, R (1980) Professional or public health?
    Nuffield Provinc. Hosp. Trust

Jarman, B (1983) Identification of underprivileged
    areas, British Medical Journal 286, 1705-8
Jarman, B (1984) Underprivileged areas, British
    Medical Journal 289, 1587-92
Jewson, N (1974) Medical knowledge and the patronage
    system in eighteenth century England, Sociology
    8, 369-85
Jewson, N (1976) The disapperance of the sick man
    from medical cosmology 1770-1870, Sociology 10,
    225-44
Jones, C (1979) Urban deprivation and the inner
    city, Croom Helm
Jones, T & M Prowle (1982) Health service finance,
    Certified Accountants Education Trust
Jordan, B (1974) Poor parents, RKP
Jordan, B (1976) Freedom and the welfare state, RKP
Jordan, B (1982) Mass unemployment and the future of
    Britain, RKP
Joseph, A E & D R Phillips (1984) Accessibility and
    utilisation, Harper and Row

Kamenka, E (1979) What is justice? in E Kamenka and
    A Tay (eds) Justice, Arnold
Kasl, S V, S Gore & S Cobb (1975) The experience of
    losing a job, Psychosomatic Medicine 37, 106-21
Khogali, M (1979) Tuberculosis among immigrants in
    the UK, Journal of Epidemiology and Community
    Health, 33
Kincaid, J C (1973) Poverty and inequality in
    Britain, Penguin
King, A (1975) Overload, Political Studies 23,
    162-74
Kirby, A M (1982) The politics of location, Methuen
Klein, R (1981) The strategy behind the Jenkin
    non-strategy, British Medical Journal 282,
    1089-91
Klein, R (1982) Reflections of an ex-AHA member,
    British Medical Journal 284, 992-4
Klein, R (1983) The politics of the national health
    service, Longman
Knox, P L (1975) Social well-being, Oxford UP
Knox, P L (1978) The intraurban ecology of primary
    medical care, Environment and Planning A 10,
    415-35
Knox, P L (1979) The accessibility of primary care
    to urban patients, Journal of the Royal College
    of General Practitioners 29, 160-68

Bibliography

Knox, P L (1982) Living in the United Kingdom, in R
    J Johnston & J C Doornkamp (eds) The changing
    geography of the UK, Methuen
Knox, P L & M Pacione (1980) Locational behaviour,
    place preferences and the inverse care law in the
    distribution of primary medical care, Geoforum
    11, 43-55

Lane, D (1976) The socialist industrial state, Allen
    & Unwin
Lange, W (1979) Marxism, liberalism and justice, in
    E Kamenka & A Tay (eds) Justice, Arnold
Lankford, P (1971) The changing location of physi-
    cians, Antipode 3, 68-72
Laski, H (1925) A grammar of politics, Allen & Unwin
Laski, H (1943) Reflections on the revolution of our
    time, Allen & Unwin
Lasswell, H D (1958) Politics, World Publishing Co
Last, J (1963) The iceberg, Lancet 2, 28-31
Laughlo, M (1984) The Spitalfields health survey,
    Department of Community Medicine, Tower Hamlets
    Health Authority
Lawless, P (1981) Britains inner cities, Harper &
    Row
Leavey, R (1982) Inequalities in urban primary care,
    Department of General Practice, University of
    Manchester
Lee, D (1959) Freedom and culture, Prentice-Hall
Lees, D S (1961) Health through choice, Institute of
    Economic Affairs
Left, S (1950) The health of the people, Victor
    Gollancz
Le Grand, J (1982) The strategy of equality, Allen &
    Unwin
Le Grand, J & R V·F Robinson (1976) The economics of
    social problems, Macmillan
Leiss, W (1978) The limits to satisfaction, Marion
    Boyars
Lenski, G (1966) Power and privilege, Free Press
Levi-Strauss, C (1970) The raw and the cooked, Cape
Lindblom, C E (1959) The science of 'muddling
    through', Public Administration 19, 79-99
Logan, R F L (1964) Studies in the spectrum of
    medical care, in G McLachlan (ed) Problems and
    progress in medical care, Oxford UP
Lukes, S (1974) Power, Macmillan

McCord, N (1978) Ratepayers and social policy, in P
    Thane (ed) The origins of British social policy
McGregor, A (1979) Area externalities and urban
    employment, in Jones op. cit.

Bibliography

McKeown, T (1979) The role of medicine, Basil
    Blackwell
McKeown, T & R G Brown (1955-6) Medical evidence
    related to population changes in the eighteenth
    century, Population Studies 9
McKinley, J (1973) Social networks, lay consulta-
    tion, and help-seeking behaviour, Social Forces
    53, 275-92
McKinley, J (1981) Social network influences on
    morbid episodes and the career of help seeking,
    in L Eisenberg and A Kleinman (eds) The relevance
    of social science to medicine, D Reidel
Manwaring, T & N Sigler (eds) Breaking the nation,
    Pluto Press
Marcuse, H (1964) One dimensional man, RKP
Marshall, T H (1965) Class, citizenship and social
    development, Doubleday Anchor
Marshall, T (1970) Social policy, Hutchinson
Marwick, A (1968) Britain in the century of total
    war
Marwick, A (1982) British society since 1945,
    Penguin
Maslow, A (1943) A theory of human motivation,
    Psychological Review 50, 370-96
Maslow, A (1954) Motivation and personality, Harper
Mather, H G, D C Morgan, & N G Pearson (1976)
    Myocardial infarction, British Medical Journal
    1, 925-9
Mauss, M (1954) The gift, RKP
Maxwell, R J (1981) Health and wealth, D C Heath
Maynard, A & A Ludbrook (1980) Budget allocation in
    the NHS, Journal of Social Policy 9, 289-312
Meacher, M (1982) Socialism with a human face, New
    Socialist March-April, 18-21
Mechanic, D (1978) Medical sociology, Free Press
Middlemas, K (1979) Politics in industrial society,
    Andre Deutsch
Midgley, M (1979) Beast and man, Methuen
Mier, R (1982) Enterprise zones, Planning 48(4),
    10-4
Miliband, R (1969) The state in capitalist society,
    Weidenfeld & Nicolson
Miliband, R (1977) Marxism and politics, Oxford UP
Miller, D (1976) Social justice, Clarendon Press
Milward, A S (1977) War, economy and society 1939-45
Ministry of Health (1945) The hospital survey of
    England and Wales (10 vols) HMSO
Ministry of Health (1962) A hospital plan for
    England and Wales, HMSO
Ministry of Health (1966) The hospital building
    programme, HMSO

Bibliography

Mishra, R (1981) Society and social policy,
    Macmillan
Mishra, R (1984) The welfare state in crisis,
    Wheatsheaf Books
Mitchell, B R & P Deane (1962) Abstract of British
    historical statistics
Mohan, J (1984a) State policies and the development
    of the hospital services in North East England,
    1948-82, Political Geography Quarterly 3, 275-95
Mohan, J (1984b) Geographical aspects of private
    hospital developments in Britain, Area 16, 191-99
Mohan, J (1985) Independent acute medical care in
    Britain, International Journal of Urban and
    Regional Research 9, 467-84
Mohan, J & K J Woods (1985) Restructuring health
    care, International Journal of Health Services
    15, 197-215
Mooney, G (1983) The NHS: caring about caring?,
    Health Economics Research Unit, University of
    Aberdeen
Morgan, K O (1984) Labour in power 1945-51,
    Clarendon Press
Morgan, M (1983) Measuring social inequality,
    Community Medicine 5, 116-24
Morgan, M & S Chinn (1983) ACORN group, social class
    and child health, Journal of Epidemiology and
    Community Health 37, 196-203
Morris, J (1975) Uses of epidemiology, Churchill
    Livingstone
Morris, J (1980) Are health services important to
    people's health? British Medical Journal 280,
    167-8
Moseley, M (1980) Rural development and its rele-
    vance to the inner city debate, SSRC Inner City
    Working Papers 9
Moser, C A, A J Fox & D R Jones (1984) Unemployment
    and mortality in the OPCS longitudinal study,
    Lancet ii, 1324-28
Moser, C A, A J Fox, D R Jones & P O Goldblatt
    (1986) Unemployment and mortality, Lancet i,
    365-7
Moser, C A & W Scott (1961) British towns, Oliver
    & Boyd

Nader, L & T Maretzki (1973) Culture, illness and
    health, American Anthrop. Assocn.
Navarro, V (1974) Medicine under capitalism, Prodist
Navarro, V (1978) Class struggle, the state and
    medicine, Martin Robertson
Nevitt, D A (1977) Demand and need, in H Heisler
    (ed) Foundations of social administration,

Bibliography

Macmillan

Nozick, R (1974) <u>Anarchy, state and utopia</u>, Basil
    Blackwell

Nuffield Provincial Hospitals Trust (NPHT) (1948)
    <u>Hospital-treated sickness amongst the people of</u>
    <u>Stirlingshire</u>, NPHT

Nuffield Provincial Hospitals Trust (1950) <u>Hospital-</u>
    <u>treated sickness amongst the people of Ayrshire</u>,
    NPHT

Nuffield Provincial Hospitals Trust (1955) <u>Studies</u>
    <u>in the function and design of hospitals</u>, Oxford
    UP

O'Connor, J (1973) <u>The fiscal crisis of the state</u>,
    St James Press

O'Dowd, L & B Rolston (1985) Bringing Hong Kong to
    Belfast? <u>International Journal of Urban and</u>
    <u>Regional Research</u> 9, 218-32

Offe, C (1974) Structural problems of the capitalist
    state, in K von Beyme (ed) <u>German political</u>
    <u>studies</u>, Sage

Offe, C (1984) <u>Contradictions of the welfare state</u>,
    Hutchinson

Offer, J (1984) Informal welfare, social work and
    the sociology of welfare, <u>British Journal of</u>
    <u>Social Work</u> 14, 545-55

OHE (Office of Health Economics) (1979) <u>Compendium</u>
    <u>of health statistics</u> (3rd Edition), OHE

Office of Health Economics (1981) <u>Compendium of</u>
    <u>health statistics</u>, 4th Edition, OHE

OHE (Office of Health Economics (1984) <u>Compendium of</u>
    <u>health statistics</u> (Fifth Edition), OHE

Office of Population Censuses and Surveys (1974)
    Commentary: social class, <u>Social Trends</u> 4

Office of Population Censuses and Surveys (1978)
    <u>Occupational mortality: decennial supplement</u>
    <u>1970-2</u>, HMSO

Office of Population Census and Surveys (1982) <u>Area</u>
    <u>mortality: decennial supplement</u>, HMSO

Office of Population Censuses and Surveys (OPCS)
    (1984) <u>General household survey 1982</u>, HMSO

Office of Population Censuses and Surveys (1985) <u>The</u>
    <u>general household survey 1983</u>, HMSO

Owen, D (1976) <u>In sickness and in health</u>, Quartet

Pacione, M (1982) The use of objective and subjec-
    tive measures of life quality in human geography,
    <u>Progress in Human Geography</u> 6, 495-514

Pahl, R E (1977) Collective consumption and the
    state in capitalist and state socialist societ-
    ies, in R Scase (ed) <u>Industrial society</u>,

Tavistock

Parker, J (1975) Social policy and citizenship,
Macmillan

Parkin, F (1971) Class inequality and political
order, Paladin

Parsons, T (1951) The social system, Free Press

Pater, J E (1981) The making of the national health
service, Kings Fund

Peacock, A & D W Wiseman (1966) The growth of public
expenditure in the UK, Allen & Unwin

Peters, R S (1958) The concept of motivation, RKP

Pilisuk, M & C Froland (1978) Kinship, social networks, social support and health, Social Science
and Medicine 12B, 273-80

Pill, R & N C H Stott (1982) Concepts of illness
causation and responsibility, Social Science and
Medicine 16, 43-52

Pinder, J (1981) Fifty years of political and
economic planning

Pinker, R (1966) English hospital statistics
1861-1938, Heinemann

Pinker, R (1971) Social theory and social policy,
Heinemann

Pinker, R (1979) The idea of welfare, Heinemann

Plant, R (1970) Social and moral theory in casework,
RKP

Plant, R, H Lesser & P Taylor-Gooby (1980) Political
philosophy and social welfare, RKP

Platt, S (1984) Unemployment and suicidal behaviour,
Social Science and Medicine 19, 93-115

Powles, J (1973) On the limitations of modern
medicine, Science, Medicine and Man 1, 1-30

Rawls, J (1972) A theory of justice, Clarendon Press

Rein, M (1970) Social policy, Random House

Rein, M (1976) Social science and public policy,
Penguin

Reiser, S J (1978) Medicine and the reign of technology, Cambridge UP

Renaud, M (1975) On the structural constraints to
state intervention in health, International
Journal of Health Services 5, 559-71

Retherford, R D (1975) The changing sex differential
in mortality, Greenwood Press

Roberts, D (1960) Victorian origins of the welfare
state, Yale UP

Roberts, D F (1976) Sex differences in disease and
mortality, in C Carter & J Peel (eds) Equality
and inequality in health, Academic Press

Robson, W A (1976) Welfare state and welfare
society, Allen & Unwin

Bibliography

Room, G (1979) The sociology of welfare, Martin
    Robertson
Rossdale, M (1965) Health in a sick society, New
    Left Review 34, 32-90
Rosser, R M & V C Watts (1974) The development of a
    classification of the symptoms of sickness and
    its use to measure the output of a hospital, in D
    Lees & S Shaw (eds) Impairment, disability and
    handicap, Heinemann
Rowntree, S (1901) Poverty, Macmillan
Rowthorn, B (1981) The politics of the AES, Marxism
    Today 25(1), 4-10
Royal College of General Practitioners (1974)
    Mortality statistics from general practice, HMSO
Runciman, W G (1966) Relative deprivation and social
    justice, RKP

Salloway, J & P Dillon (1973) A comparison of
    family networks and friend networks in health
    care utilisation, Journal of Comparative Family
    Studies 4, 131-42
Samuelson, P A (1973) Economics, McGraw-Hill
Sargeant, T (1979) Joint care planning in health and
    personal social services, in T A Booth (ed)
    Planning for welfare, Basil Blackwell & Martin
    Robertson
Scott-Samuel, A (1977) Social area analysis in
    community medicine, British Journal of Preventive
    and Social Medicine, 31, 199-204
Scott-Samuel, A (1984) Need for primary health care,
    British Medical Journal 288, 457-8
Scrambler, A, G Scrambler & D Craig (1981) Kinship
    and friendship networks and women's demand for
    primary care, Journal of the Royal College of
    General Practitioners 26, 746-50
Seabrook, J (1982) Unemployment, Paladin
Seabrook, J (1983) The idea of neighbourhood, Pluto
    Press
Shannon, G W & G E A Dever (1974) Health care
    delivery, McGraw-Hill
Shaw, J M (1979) Rural deprivation and social plan-
    ning, in J M Shaw (ed) Rural deprivation and
    planning, Geobooks
Silman, A J & S Evans (1981) Regional differences in
    survival from cancer, Community Medicine, 3,
    291-97
Simmel, G (1950) The sociology of Georg Simmel, Free
    Press
Social Services Committee (1985) Community care with
    special reference to adult mentally ill and
    mentally handicapped people (3 vols), HMSO

Bibliography

Spoehr, A (1956) Cultural differences in the inter
    pretation of natural resources, in W L Thomas
    (ed) Man's role in changing the face of the
    earth, University of Chicago Press
Smith, D M (1973) The geography of social well-being
    in the US, McGraw-Hill
Smith, D M (1974) Who gets what where and how,
    Geography 59, 289-97
Smith, D M (1977) Human geography, Arnold
Smith, D M (1979) Where the grass is greener,
    Penguin
Smith, F B (1979) The people's health 1830-1910,
    Croom Helm
Smith, R (1985) Occupational health, British Medical
    Journal 291, 1024-7, 1107-11, 1191-5, 1263-6,
    1338-41, 1409-12, 1492-5, 1563-6, 1626-9, 1707-10
Stafford, E M, P R Jackson & M H Banks (1980)
    Employment, work involvement and mental health in
    less qualified young people, Journal of Occupa-
    tional Psychology, 53, 291-304
Stevens, R (1966) Medical practice in modern
    England, Yale UP
Stevenson, J (1984) British society 1914-45, Penguin
Stretton, H (1976) Capitalism, socialism and the
    environment, Oxford UP
Suchman, E A (1964) Sociomedical variations among
    ethnic groups, American Journal of Sociology 70,
    319-31
Susser, I (1983) Norman Street, Oxford UP

Tawney, R H (1938) Religion and the rise of
    capitalism, Penguin
Tawney, R H (1961) The acquisitive society, Fontana
Tawney, R H (1964) The radical tradition, Penguin
Taylor, R (1979) Medicine out of control, Sun Books
Taylor-Gooby, P & J Dale (1981) Social theory and
    social welfare, Arnold
Thomas, C & S Winyard (1979) Rural incomes, in J M
    Shaw (ed) Rural deprivation and planning,
    Geobooks
Thomas, H E (1968) Tuberculosis in Britain,
    Proceedings of the Royal Society of Medicine 61
Thomas, L (1975) The health care system, New England
    Journal of Medicine 293, 1245
Thompson, E (1971) The moral economy of the English
    crowd in the eighteenth century, Past and Present
    50
Thunhurst, C (1985) The analysis of small area
    statistics and planning for health, The
    Statistician 34, 93-106
Thurow, L C (1981) The zero-sum society, Penguin

Bibliography

Titmuss, R M (1965) Income distribution and social change, Allen & Unwin
Titmuss, R M (1968) Commitment to welfare, Allen & Unwin
Titmuss, R M (1973) The gift relationship, Penguin
Titmuss, R M (1974) Social policy, Allen & Unwin
Titmuss, R M (1976) Problems of social policy, HMSO
Toland, S (1980) Changes in living standards since the 1950s, Social Trends 10, 13-38
Totman, R (1979) Social causes of illness, Pantheon Books
Townsend, P (1954) Measuring poverty, British Journal of Sociology 5, 130-37
Townsend, P (1976) The difficulties of policies based on the concept of area deprivation, Queen Mary College, Department of Economics, Barnett Shine Foundation Lecture
Townsend, P (1979) Poverty in the UK, Penguin
Townsend, P (1980) Social planning and the Treasury, in N Bosanquet & P Townsend (eds) Labour and equality, Heinemann
Townsend, P & N Davidson (1982) Inequalities in health, Penguin
Tuckett, D (1985) Meeting the experts, Tavistock
Turnbull, C (1973) The forest people, Paladin
Turshen, M (1977) The British national health service, Health Service Action

Wainwright, H & D Elliot (1982) The Lucas plan, Allison & Busby
Walker, A (1984) Social planning, Basil Blackwell & Martin Robertson
Wallman, S (1984) Eight London households, Tavistock
Walters, V (1980) Class inequality and health care, Croom Helm
Ward, M (forthcoming) Attitudes to and mobility in rural housing in Norfolk, University of London, PhD Thesis
Watkin, B (1978) The national health service: the first phase, Allen & Unwin
Watkins, S J (1984) Recession and health - the policy implications, in G Westcott (ed) Health policy implications of unemployment, WHO
Westergaard, J & H Reser (1976) Class in a capitalist society, Penguin
Wheeler, M (1972) How many acute beds do we really need? British Medical Journal 4, 220
Widgery, D (1979) Health in danger, Papermac
Wilcocks, A J (1967) The creation of the national health service RKP
Wilkinson, R G (1973) Poverty and progress, Methuen

Bibliography

Williams, A (1974a) 'Need' as a demand concept? in A
    J Culyer (ed) Economic policies and social goals,
    Martin Robertson
Williams, A (1974b) Measuring the effectiveness of
    health care systems, British Journal of
    Preventive and Social Medicine 28, 196-202
Williams, R (1971) The country and the city, Chatto
    & Windus
Williams, R (1977) Marxism and literature, Oxford UP
Williams, R (1983) Concepts of health, British
    Journal of Sociology 17, 185-205
Wilson, E (1980) Marxism and the welfare state, New
    Left Review 122, 79-89
Wilson, E O (1975) Sociobiology, Harvard UP
Winkler, J (1976) Corporatism, European Journal of
    Sociology 17, 100-36
Wollheim, R (1976) Need, desire and moral turpitude,
    Macmillan
Woods, K J (1982) Social deprivation and resource
    allocation in the Thames regional health
    authorities, in Health Research Group (ed)
    Contemporary perspectives on health and health
    care, Queen Mary College, Department of Geography
    and Earth Science, Occasional Paper 20
Woodward, J (1974) To do the sick no harm, Arnold
World Health Organisation (WHO) (1961) Constitution,
    WHO

Zola, I K (1972) Medicine as an Institution of
    social control, Sociological Review 20, 487-504

Index